CHARGE!

The Napoleonic Library

Other books in the series include:

1815: THE RETURN OF NAPOLEON
Paul Britten Austin

DECLINE AND FALL OF NAPOLEON'S EMPIRE
How the Emperor Self-Destructed
Digby Smith

LIFE IN NAPOLEON'S ARMY
The Memoirs of Captain Elzéar Blaze
Introduction by Philip Haythornthwaite

THE MEMOIRS OF BARON VON MÜFFLING
A Prussian Officer in the Napoleonic Wars
Baron von Müffling

WATERLOO LECTURES
A Study of the Campaign of 1815
Colonel Charles Chesney

WATERLOO LETTERS
A Collection of Accounts From Survivors of the Campaign of 1815
Edited by Major-General H. T. Siborne

www.frontline-books.com/napoleoniclibrary

CHARGE!

GREAT CAVALRY CHARGES OF THE NAPOLEONIC WARS

Digby Smith

Frontline Books

Charge! Great Cavalry Charges of the Napoleonic Wars

A Greenhill Book

First published in 2003 by Greenhill Books, Lionel Leventhal Limited
www.greenhillbooks.com

This edition published in 2015 by

Frontline Books
an imprint of Pen & Sword Books Ltd,
47 Church Street, Barnsley, S. Yorkshire, S70 2AS
For more information on our books, please visit
www.frontline-books.com, email info@frontline-books.com
or write to us at the above address.

Copyright © Digby Smith, 2003

The rights of Digby Smith to be identified as the author of this work has been asserted by him in accordance with the Copyright, Designs and Patents Act 1988.

ISBN: 978-1-84832-819-8

All rights reserved. No part of this publication may be reproduced, stored in or introduced into a retrieval system, or transmitted, in any form, or by any means (electronic, mechanical, photocopying, recording or otherwise) without the prior written permission of the publisher. Any person who does any unauthorized act in relation to this publication may be liable to criminal prosecution and civil claims for damages.

CIP data records for this title are available from the British Library

Printed and bound by CPI Group (UK) Ltd, Croydon, CR0 4YY

Contents

	Acknowledgements	6
	List of Illustrations	7
	List of Maps	8
	Introduction	9
Chapter 1	Types of Cavalry	11
Chapter 2	The Battle of Marengo	24
Chapter 3	The Battle of Austerlitz	46
Chapter 4	The Battle of Preussisch-Eylau	61
Chapter 5	The Battle of Albuera	76
Chapter 6	The Clash at García Hernandez	113
Chapter 7	The Battle of Borodino	122
	The Russian Raid on the Northern Flank 122	
	The Storm of the Grand Battery 129	
Chapter 8	The Battle of the Beresina Crossing	146
Chapter 9	The Clash at Haynau	157
Chapter 10	The Clash at Liebertwolkwitz	164
Chapter 11	The Battle of Möckern	183
Chapter 12	Allied Cavalry Raids of 1813	194
Chapter 13	The Battle of Fère-Champenoise	207
Chapter 14	The Battle of Waterloo	222
	The Charge of the Guards and Union Brigades 227	
	The Grand Failure 232	
	Appendices	242
	Index of Persons	300

Acknowledgements

Many people have been kind enough to share their information with me; they have contributed to the accuracy and interest of this book. Among them I should like to thank Luis Sorando Muzas for his fascinating snippets on the trophies taken in the Battle of Albuera, Marc Moerman, John Cook, Kevin Kiley, George Nafziger and Jonathan North. Finally, all those participants in the internet forums from whom I have learned so much in the last few years.

Digby Smith.

List of Illustrations

1. Kellermann's charge at Marengo page 97
2. Bessières and the Consular Guard at Marengo page 97
3. General Melas page 98
4. General Zach page 98
5. General Desaix page 98
6. Grenadiers-à-Cheval at Austerlitz page 99
7. Mamelukes at Austerlitz page 99
8. Freiherr von Kienmayer page 100
9. Prince Johann von Liechtenstein page 100
10. General Rapp with Austerlitz trophies page 100
11. Napoleon at Eylau page 101
12. Eylau battle scene page 101
13. Marshal Davout page 102
14. Marshal Soult page 102
15. Count Platov page 102
16. General Bennigsen page 102
17. General Uvarov during his raid at Borodino page 103
18. Russian cuirassiers at Borodino page 103
19. Monument to Uvarov's raid page 104
20. Borodino Church page 104
21. Crossing the Beresina page 105
22. Don Cossacks page 105
23. Blücher's cavalry at Haynau page 106
24. Prince Eugène page 106
25. Feldmarschall Blücher page 107

26. French hussars at Leipzig — page 107
27. Battle of Leipzig — page 108
28. Emperor Franz I of Austria — page 108
29. King Frederick William III of Prussia — page 108
30. Field Marshal Schwarzenberg — page 109
31. General Langenau — page 109
32. Prussian cavalry at Möckern — page 109
33. Cossacks, Bashkirs and Kalmucks — page 110
34. Russian hussars — page 110
35. French Young Guard — page 111
36. General Colbert and Guard Lancers — page 111
37. 2nd Life Guards at Waterloo — page 112
38. Inniskilling Dragoons at Waterloo — page 112

List of Maps

Battle of Marengo — page 33
Battle of Austerlitz — pages 54–55
Battle of Eylau — pages 68–69
Battle of Albuera — page 79
Battle of Borodino – The Northern Flank — page 127
Battle of Borodino – The Final Assault on the Raevski Battery — page 131
Passage of the Beresina — pages 148–149
Ambush at Haynau — page 160
Liebertwolkwitz — page 169
Battle of Möckern — page 184
Battle of Fère-Champenoise — page 217
Battle of Waterloo – Crisis of the Battle — pages 234–235

Introduction

The characteristics of the modern armoured 'cavalry' on the battlefield are: protection, mobility, shock action and firepower. These characteristics were as valid in the Napoleonic era as they are today. Cavalrymen in 1800 did not always have the protection of armour, as is enjoyed by modern tank troops, but they certainly had increased mobility compared with the infantryman. The power of their swords, and the brute force of a squadron of mounted men at speed, gave them the historical equivalent of the shock of the modern tank.

The rules of the game today are that 'cavalry' may be used to take ground, but should not be expected to hold it on their own. The Battle of Borodino will give us an example of what happens to cavalry when it is given such an inappropriate task. To succeed at this defensive task today's 'cavalry' need the close support of infantry, artillery, engineers and air power. Ideally, cavalry should always be employed – offensively and defensively – in close integration with the other arms if the best is to be made of its potential. These principles applied also to the cavalry of 1792–1815.

It is perhaps not widely realised today just how the mechanics of a remount service in the days of cavalry worked. Whereas tanks and lorries can be built and churned off the assembly line more or less as required, the 'production' of cavalry mounts was much more difficult. This was a complex procedure. A mare was usually not first covered until she was at least three years old and gestation in the horse lasts 310–370 days, call it a year. The vast majority of mares drop only one foal for each pregnancy and that foal will not be fit for military service until it is at least five years old when it has to be given specific training if it is to be useful in battle. The 'lead time' for producing new horses to replace the thousands lost, for instance, by the French in Russia meant that there would be a great shortage (and their cost would shoot through the roof if market forces applied) for some years to come. The fruits of the 1813 breeding programme would not be ready for harvest until 1819 at the earliest.

Concerning the content of this book, I was sure that there was little merit in just trotting out a series of descriptions of cavalry charges. This would be – at best – repetitive and boring. One was very much like another (in essence) once the order to go had been given. I have thus devoted considerable space to explaining the strategic environment in which each of these actions took place, so that the overall context of each situation may be understood. Not all the actions selected were simple charges at the centre of major battlefields. This too is deliberate, the aim being to demonstrate other tasks of which cavalry was capable, such as scouting, screening and long range penetration of enemy rear areas. I have included aspects from some of the great battles, but space has also been devoted to some less well-known but equally dramatic actions, such as that of Fère-Champenoise in 1814. Finally, an effort has been made to include actions involving cavalry forces other than the familiar French and British units most often discussed in English-language books. This has involved the use of sources other than the well-known English and French works. The information gleaned from them, and from several very productive debates on the internet, has produced a selection of cameos which may challenge popularly accepted versions of events. To enable the interested reader to research further, the bibliography is split up and inserted where relevant, at the end of the description of each action.

Chapter 1

Types of Cavalry

Heavy Cavalry

By 1792 heavy, armoured cavalry, called *cuirassiers* in French and *Kürassiers* in German, were the exception in European armies. Cuirassiers (as they are usually also known in English) were the descendants of the old armoured knights of the medieval period. Over time, the mutually exclusive priorities of mobility and protection – though mobility itself is protection – had led by the 16th century to the abandonment of all metal body protection for heavy cavalry, with the exception of a helmet and a breastplate. This dramatically reduced the weight of the rider on the battlefield and meant that smaller horses could be used than hitherto. Smaller horses were more readily available than the monstrous Shires and similar breeds previously needed, were faster than their larger comrades, cheaper to buy, and they ate less fodder, thus costing less to feed. Even so, these regiments were expensive to equip and maintain, relative to the other types of cavalry; there were thus few of them, though Austria, France, Russia and Prussia all had a significant cuirassier force.

Cuirassiers were generally armed with a heavy, straight-bladed sword and a pair of pistols. Carbines were not normally carried. Their armour conferred some degree of protection from small-arms' fire as well as in hand-to-hand combat. In the French Army all breastplates had to pass an acceptance test before they were taken into service. A pistol would be discharged at close range at the armour; if the ball penetrated, the item would be rejected. The cuirassiers' role in battle was to be used in compact masses, for shock action. Napoleon was a master of this tactic. Cuirassiers were not used for scouting, skirmishing or patrol duties.

The first armoured cavalry regiments of the Austrian Army appeared in 1618 and by the time of the French revolution there were nine in existence. In 1798 the two existing Karabinier regiments were converted to Kürassiers and a new regiment was raised, bringing the total of armoured cavalry regiments to 12. These were reduced to eight in 1802 and this number remained in service for the rest of the Napoleonic Wars.

In 1665 France had one armoured cavalry regiment, and this remained the case until 1803. This regiment was entitled the Cuirassiers du Roi up to 1792 and was ranked as number 8 in the seniority list of horse regiments, first established in 1635. It retained this number throughout the Napoleonic period. By 1779 there were 22 regiments of horse, and five more were raised in 1791 and 1792. By 1796 these had been reduced to 25. By 1803 the number had been reduced to 12, and in 1803 Napoleon converted these regiments to cuirassiers proper, wearing helmets and armoured breast and back plates. A 13th Regiment was added in 1808 and a 14th two years later, from Dutch service. Marshal Davout raised another regiment of this arm in Hamburg in 1813, but the men and the horses were absolutely raw and the unit soon faded into history before the siege of that city was over. This '15th Cuirassiers' was an unofficial unit and was not included in the army list. The two regiments of Carabiniers in the French Army also wore body armour from 1810.

The Russian Army gained its first cuirassiers in 1731; by 1786 there were four regiments and in 1813–14 there were 13, including four in the Guards.

Prussia had 13 nominal Kürassier regiments by the time of the French Revolution, but they wore no armour for most of the Napoleonic period, this being withdrawn in 1790. These regiments were destroyed in 1806. Four new regiments were raised in 1808, and these were given armour in 1814.

In 1695 Saxony had seven Kürassier regiments, who wore only armoured breastplates. By 1810 the order of battle had been reduced to the Garde du Corps and two line regiments. The army of the Kingdom of Westphalia (1807–13) had two armoured regiments – much to Napoleon's chagrin since he thought them too expensive. The Grand Duchy of Warsaw (1807–13) had one armoured regiment – its 14th – and King Joseph raised one in Spain in 1812.

Britain had no armoured cavalry throughout the period, and neither had Denmark, Portugal, Spain, Sweden, or the Italian states.

Line Cavalry
This category includes the various types of regiments known as horse grenadiers, carabiniers, dragoon guards and dragoons. Their principal role was shock action on the battlefield. Various armies also had regiments described as light dragoons or light horse. These were light cavalry, riding smaller horses than dragoons proper and operating in a similar way to the hussars and chasseurs-à-cheval described below.

Although the distinction between dragoons and all the other types of line cavalry was being lost by the Napoleonic period, this had not always been the case. Originally, these regiments were simply composed of mounted infantry; their horsemanship was rudimentary and their horses of lower quality than those of the 'real' cavalry regiments. There is some mystery as to how the dragoons got their name. One theory is that they were named after the large pistols the first regiments carried, the muzzles of which were in the shape of a dragon's head. In Napoleonic times they were still armed with muskets and supposedly trained to fight on foot. Indeed several French dragoon regiments were used as infantry during the Peninsular War in Spain and other theatres. Instead of the usual trumpeters, they often had drummers to give their tactical signals. Dragoons were usually armed with straight swords, short infantry muskets and a pair of pistols.

The variations of horse grenadiers and carabiniers were tactically much the same as dragoons in the Revolutionary and Napoleonic period (except in France where the carabiniers wore cuirasses in the later years as already noted). Horse grenadiers were also an early variety of dragoons, originally trained to fight on foot. Carabiniers was the title commonly given to a horse regiment when it was equipped with a short musket or carbine. By the 1790s these titles had no relation to the actual role of the regiments concerned.

Most French line cavalry were dragoons. Dragoons were some of the oldest cavalry regiments in the French Army, dating back to before 1656. In 1761 there were 16 regiments. By 1792 there were 18 regiments and three more were raised in 1793. Under Napoleon these increased to one regiment in the Imperial Guard and 30 in the line. As well as dragoons, the French Army had carabiniers and horse grenadiers. The 1st Carabiniers was raised in 1693; in 1788 this regiment was split to form the two that we see throughout the Revolutionary and Napoleonic era. There was also the famous regiment of Grenadiers-à-Cheval in the Imperial Guard.

Prussia had dragoon regiments as far back as 1675. At the time of the death of Frederick the Great in 1786 there were 12 regiments and two more were added by 1806. When the Prussian Army was reorganised after the disasters of Jena and Auerstädt in that year, six regiments were formed and in 1815 two more were added.

Russia introduced dragoons under Peter the Great (1672–1725) and had 37 regiments in 1803, including the Life Guards Regiment. After 1813 only 19 of these remained; eight were converted to mounted rifle regiments in that year, the rest to lancers.

The British Household Cavalry of this period included the 1st and 2nd Life Guards and the Royal Horse Guards. In 1690 this latter regiment was nicknamed 'The Blues' from the colour of its coat. The line cavalry included regiments of dragoon guards and dragoons, though these designations had very limited practical effects on the equipment or tactical employment of the regiments. Only the uniforms changed. The dragoon guards were the descendants of regiments formerly known simply as 'horse'. There were seven regiments of dragoon guards as well as six of dragoons (up to 1799, five thereafter) during the Napoleonic era. The light dragoon regiments were numbered in the same sequence as the dragoon regiments, though their tactical role was normally different.

In 1700 Austria had various dragoon and light horse

regiments and there was much chopping and changing between these arms throughout the 18th century, although the only real difference between them seemed to be in the colour of their coats. In 1798 the then six existing dragoon regiments were converted into light dragoons as were the seven 'Chevauxlégers' regiments. At the same time, two more light dragoon regiments were raised, making 15 in all. In 1802 five of these regiments were converted back to dragoons, and six were remustered as Chevauxlégers. The other four light dragoon regiments were disbanded. One Kürassier regiment was converted to dragoons thus making a total of six. These regiments were retained through the rest of the Napoleonic period.

Light Cavalry
This category included units variously named as hussars, chasseurs-à-cheval and Jäger-zu-Pferde (both sometimes translated as mounted rifles), light dragoons, and light horse, according to which was their parent army. The regiments of these light cavalry arms are grouped together because the only perceptible difference between them was often their titles. The costumes worn by these arms were also often barely distinguishable one from another.

The name hussar was originally *cursarius* in the Middle Ages; it came from the old Italian and meant 'corsair, robber or pirate'. The title 'mounted rifles' is also a misnomer since only a select few in each regiment would generally be armed with carbines and few of these would be rifled weapons. These regiments were mounted on small horses, almost ponies, and were mostly used for scouting and patrol duties, screening their own army from enemy observation and bringing in intelligence on hostile movements, capturing despatches and couriers. They were rarely used for shock action. Their arms were usually a light curved sabre, a carbine and a pair of pistols.

Hussars originated in Hungary and variations on the braided Hungarian national costume were adopted by hussar regiments in other armies. The pelisse (*Pelz*) was originally a wolfskin, worn over the left shoulder to protect against sabre cuts. Hussars were part of the original Austro-Hungarian Army; from there they spread and were copied in almost all European armies by 1792, although they had died out again in the armies of some of minor states by then.

In 1734 Austria had three regiments of these Hungarian light horse; by 1762 this had risen to 15, and by 1812 there were 12.

France raised its first hussar regiment, the Hussards royaux, in 1692, and by 1791, after various changes, there were six regiments. In 1803-10 there were ten regiments; the annexation of Holland in 1810 raised the total to 11; three more were raised in 1813, and one was raised and disbanded again in 1814.

The first six regiments of chasseurs-à-cheval were raised in the French Army in 1779 and another six were added in 1780. Thirteen more regiments were raised during 1793-95, making 25 in all. The 17th was disbanded in 1795 and this number remained vacant thereafter. A 26th Regiment was raised in 1802; the 27th-29th in 1808, the 31st in 1811. The number 30 remained vacant. The Imperial Guard had a regiment of chasseurs-à-cheval, one of Napoleon's favourites.

Hussars made their first appearance in the Prussian Army in 1721. Ten years later there were two hussar regiments and during the reign of Frederick the Great this increased to ten, although the 9th Regiment was the Towarczys and Bosniaken, a regiment equipped with lances. In 1806 11 hussar regiments (including the Towarczys and Bosniaken) took the field. After the catastrophic Prussian defeat in that year, six new regiments were raised for the reformed Prussian Army in 1808 and by the time of the Waterloo campaign there were 12 hussar regiments in the Prussian Army.

The first hussar regiments in the Russian Army were raised on 14 October 1741 by Princess Anna Leopoldovna. They were named after the nations from which they came: Gruzinskiy (Georgia), Moldavskiy (Moldavia), Serbskiy (Serbia) and Vengerskiy (Hungary). By 1761 there were five regiments, and by 1765 these had increased to 12. In 1776 these were increased to 15, but during 1783-96 there were no hussars in the regular Russian Army at all. The arm was re-introduced at a strength of one guard and eight line regiments by Czar Paul I in 1796. In 1811 there were 11 regiments in the line plus the Guard Hussars, but in that year, the hussars were also armed with lances. Mounted rifle regiments appeared first in Russia in the 1780s.

Saxony raised its first hussar regiment in 1791 and this was the only regiment of its kind in the army through the Napoleonic era.

Britain's regular army was late in introducing hussars, although the yeomanry had fallen in love with the showy costumes by 1792. In 1812 the 7th, 10th, 15th and 18th Light Dragoons were renamed as hussars, as were the three light dragoon regiments of the King's German Legion. This was the extent of British hussar regiments in the Napoleonic era.

Lancers

In most armies lancers were light horse units, generally used for the same roles as the hussars or other light cavalry. In the English-speaking world, these regiments were named after the weapons they bore but in French they were described as chevaux-légers (light horse) or chevaux-légers-lanciers.

The eastern European original name for this cavalry arm was Oghlan. It was first used in the Crimea in the period 1441–1683, when the lifeguards of the Khans of the Tartars in that region were armed with these weapons. Only the front rank of each regiment carried lances, and these were men of the minor nobility of the khanate, the Oghlans. The second rank was made up of the squires of the Oghlans; these men carried only sabres but in later years had carbines as well. The old kingdom of Poland stretched into the Tartar areas and the Polish army came to contain many regiments of this type. The name Oghlan gradually became corrupted to 'Ulan' or 'Uhlan'.

In any mêlée with opposing cavalry, the front rank of lancers would have a distinct advantage in the initial shock; thereafter, the lance could be a hindrance. Against infantry, the lancers would seem to have had a major edge, particularly if the infantry were unable to fire their weapons, as will be shown in the chapter on Albuera. Despite this, instances of lancers breaking squares are almost unknown.

As with the hussar regiments who wore Hungarian costume, lancer regiments in all armies of the period adopted the Polish national costume; the square-topped *czapka* headdress, the *kurtka* jacket and the characteristic waist sash with the 'waterfall' in the small of the back.

The first lancer regiment appeared in the French Army in 1734 under the Marshal de Saxe, during the War of the Polish Succession. In true Polish fashion, the first rank were members of the minor nobility and their squires (*pacholken*) formed the second rank. This regiment was disbanded in 1750 and there followed a long period in which the French Army had no lancers. Napoleon showed his interest in the lance as early as 1809, when the 1st (Polish) Chevaux-légers of the Imperial Guard were equipped with that weapon. When Holland was annexed in 1810, a 2nd (Dutch) Chevaux-légers-lanciers Regiment was raised in the Guard, and in 1812 there was – briefly – a 3rd (Lithuanian) Regiment. It was raised in the summer in Warsaw and Vilna and largely destroyed by the Russians at Slonim on 19 October of that year.

Polish troops in French service included the Legion of the Danube, raised in September 1799; it included a regiment of lancers. On 31 March 1808 the legion became the Legion of the Vistula, which had two such regiments. Following the dramatic success of the 1st Lancers of the Vistula in the Battle of Albuera, on 18 June 1811 Napoleon decreed the conversion of six dragoon regiments and the cavalry of the Legion Hanovrienne to lancers. There were thus nine lancer regiments of the line in the French Army until 1814. The 7th and 8th were not re-raised in 1815, but the lancers of the Vistula Legion formed the new 7th.

Prussia raised its first Ulan regiment in 1741, but it was converted to hussars in the next year. Lancers popped up again in 1745 when a regiment of 'Bosniaken' was raised. Following Prussia's annexation of areas of the defunct Polish republic in 1795, the Prussian Army took a regiment of Tartars into service; in 1800 this unit was retitled 'Towarczys' (comrades). These two units were ranked as the 9th Regiment in the hussars. In 1807 these men were used to form the 1st and 2nd Ulans, a 3rd Regiment was raised in 1809, and in 1815 the 4th–8th followed.

The first three Ulanen regiments in the Saxon army were raised in 1754, but disbanded again shortly afterwards. In 1813 a regiment of Chevaulégers was converted to lancers.

Ulans first appeared in the Austrian Army in 1784, when an 'Ulanen-Pulk' or regiment of 300 Towarziken (comrades) and 300 Pozdonen (squires) was raised in the provinces of Galicia and

Lodomerian, which had been taken from Poland in 1772. In 1798 this corps became the Merveldt Ulans (Nr 1). In 1790 Graf O'Donell's *Freikorps* was raised in Galicia and it included four squadrons of Ulans, which became an independent regiment in the next year. In 1793 it became the 2nd Ulan Regiment (Schwarzenberg). The 3rd Ulans followed in 1801, and in 1805 Prinz Carl became their colonel-in-chief. In 1813 the Kaiser Ulans (Nr 4) completed the series for the Napoleonic period.

Strangely enough, even though Russia had swallowed much of the old Kingdom of Poland by 1795, it was not until 1802 that a lancer regiment appeared in the Russian Army. By 1812 there were the Guard lancers and six such regiments in the line, but by then even the 11 hussar regiments as well as the numerous Cossack *pulks* bore lances. In 1813 11 regiments of dragoons were converted to lancers.

The cavalry of the Grand Duchy of Warsaw (1807–13) contained no fewer than 15 lancer regiments. On 1 July 1812 Napoleon proclaimed the creation of the Duchy of Lithuania; in its very brief existence its army contained four cavalry regiments – all lancers – numbered 17th–20th in the seniority list of the Grand Duchy of Warsaw. The Lithuanian Gendarmes were also armed with the lance and charged with Doumerc's cuirassiers at the Beresina. The lance long remained a popular weapon with the Polish forces – at the outbreak of the Second World War, the Polish Army had no fewer than 27 regiments equipped with the lance.

Despite the popularity of lancers in continental armies, it was not until 1816 that six British cavalry regiments were equipped with this weapon.

Irregular Cavalry
Formations of this type in the Napoleonic period include the many Cossack *pulks*, the Bashkirs, Tartars and Kalmucks of the Russian Army in 1812. Such loosely organised, partially trained and disciplined regiments, mounted on light, sturdy ponies and armed with sabres, lances – and even bows and arrows – were of no use in shock actions on the battlefield. Such irregular bands were, however, excellent for scouting missions, capturing couriers and despatches, or striking at stragglers and supply convoys.

They were also quite effective in guerrilla-type raids in enemy flank and rear areas, as was often demonstrated in the later stages of the 1812 campaign and in 1813 in western Germany. During the early part of this campaign, the Cossack commander, Colonel Tettenborn, with a small force, actually 'captured' the city of Hamburg on 17 March and held it for a few weeks before being ejected again by Davout. Later, on 30 September, General Czernichev captured King Jérôme's capital city of Kassel with a raiding force of 1,300 Cossacks, dragoons and hussars and held it until 3 October, throwing the administration of the Kingdom of Westphalia into panic.

Cavalry Tactics

Cavalry tactics may be defined as the day-to-day use of mounted troops on the battlefield. They were – and still are – dictated by the capabilities and limitations of the components of the mounted arm: man, horse, weapons and equipment. The various evolutions and formations which a nation's cavalry should use were often worked out and written down at regimental level in the early years. By the Napoleonic era these had been standardised at army level and were republished from time to time as changes in weaponry, tactical thought or battlefield experience dictated. As all nations used the same beasts and (mostly) the same weapons, there was great commonality between these publications as use of cavalry had matured on the battlefields of the Seven Years War. The exception to this was the spread – or reintroduction – of the use of the lance as has been mentioned above. This change in cavalry tactics and weapons was mirrored in the infantry by the introduction of light troops and rifles following the Seven Years War and the War of American Independence.

A cavalry regiment at this time usually consisted of from three to five squadrons, each squadron about a hundred men strong, although this number varied from nation to nation. The squadron was usually divided into two troops. Tactical signals were given by trumpet calls. When deployed for action, the squadron formed in two ranks, with a 12-pace interval to the next squadron. From this line, a regiment could advance as a whole, or lead off from the right or left flank 'in echelon'. If two regiments were present on a narrow frontage, the second regiment would maintain a distance of about 100 paces behind the leading unit. In this formation, it was usual for the inter-squadron spaces to be the width of an entire squadron. The following regiment would align itself so that its squadrons were in the spaces left between the squadrons in the front unit. This allowed the front regiment to fall back through the following unit if repelled. This formation was known as 'en echiquier'.

French line cavalry was drilled according to the *Ordonnance* of 1788; for light cavalry, a separate *Instruction Concernant l'Exercise et les Manoeuvres de la Cavalerie Légère* was issued in 1799. Both regulations were replaced by the 1804 *Ordonnance Provisoire sur l'Exercise et les Manoeuvres de la Cavalerie*. On 24 September 1811 a drill manual for the newly converted lancer regiments, *Instruction sur l'Exercise et les Manoeuvres de la Lance*, was published.

Contrary to Hollywood myths, cavalry regiments did not always move at a gallop. This would have exhausted even well fed, healthy horses within a very short time. Charges started off at a walk and only progressed through a trot to a gallop when within about 200 paces of the enemy line. The rear rank of troopers would keep about three paces from the rumps of the horses in the front line until about 50 paces from the target. They would then close this gap and fill up any spaces in the front rank left by casualties, in order to maximise the shock of their impact on the target. If there were a second regiment following the first, it would keep about 250–300 paces behind and would never move at more than a trot. Should enemy cavalry be overthrown in a charge, the victorious commander would (according to the book) have the recall sounded and advance slowly, gathering in any scattered troopers and reforming their ranks before re-engaging the enemy. When attacking artillery frontally, the inter-file distances would often be deliberately increased to reduce the effects of the inevitable discharge of canister.

British cavalry regiments earned a reputation in the Peninsula for 'galloping at everything', the officers losing any control over their men and the whole event resembling a steeplechase. Several great tactical opportunities were squandered in this manner; Wellington was most scathing of his cavalry commanders.

Any regiment that had executed a charge would then need to rest its horses to allow them to recover. The more strenuous the charge, the longer the rest that would have to be granted. Violent actions, with repeated charges, would quickly result in such 'charges' being carried out at a walk. There was also the need to allow the horses to drink periodically and to feed. This limited the duration for which cavalry could be employed without their performance being drastically reduced.

Horses that are required to do physical work have to be properly fed. In western Europe during the Napoleonic period the key ingredient for maintaining the performance of a horse was oats. The wrong diet will quickly kill a horse with colic; it was wrong diet – no oats, too much green fodder – which wiped out much of Napoleon's cavalry in the early stages of his invasion of Russia in 1812. For some reason, the Cossacks' sturdy, primitive and undemanding ponies seemed to be able to maintain their work performance on just hay, grass and anything else that came by.

Control of a squadron or regiment in the deafening hell of a battlefield was vital if it were to achieve its mission. Obviously, the human voice was inadequate for this task, and kettledrums had been abandoned as being too heavy and bulky for the purpose. The trumpet was small, light, robust and loud enough to fit the bill. By the 1790s most armies had a standardised set of trumpet calls, which would tell the men of a regiment to put their horses into the various gaits, to change direction, halt, retire and so on. Each regiment would precede its call with its own 'signature tune' – one or two distinctive bars, which all members of the regiment would know – so that general chaos in the cavalry on the field would be avoided. The trumpeter was thus a key member of the commander's team. Usually, there were one or two trumpeters to each squadron, and another at regimental level.

In order to make them easy for the commander to find in the confusion of the battlefield, trumpeters were normally dressed in distinctive uniforms. For British regiments, and for those of several other nations, if the regiment wore blue coats faced in red, the trumpeter would wear a red coat faced in blue, reversed colours as they were called. Other nations covered the sleeves of

Types of Cavalry

trumpeters' tunics with lace chevrons and topped the sleeves with 'swallows nests', usually in red, edged and covered with lace braid. Most of them wore red plumes or crests to their helmets. In almost all armies of the period, trumpeters were mounted on greys though it seems that this was not the case in Russia until 1811, and trumpeters in Prussian Kürassier regiments did not ride greys.

CHAPTER 2

The Battle of Marengo
14 June 1800

That this battle should have contributed so much to the legend of Napoleon's invincibility is a mystery to many. As we shall see, Napoleon blundered very badly in the days leading up to this action only for his chestnuts to be pulled from the fire by his friends, Desaix and Kellermann, at the very last minute. Following Napoleon's brilliant campaign of 1796, and his dictated Peace of Campo Formio of 17 October 1797, the Austrian presence in Italy had been all but wiped out. A crop of republics – on the French model and firmly under French control – had sprung up in its place. Piedmont, indeed, had been incorporated into metropolitan France. In March 1798, the Directory – the French government – engineered another revolution in the Swiss Confederation; this became the Helvetian Republic, which, in August of that year, entered into an alliance with France, thus giving France control of all the strategic Alpine passes linking Germany with Italy.

In May 1798 Napoleon had set off on his historic expedition to Egypt, in which place he was effectively marooned after Admiral Nelson destroyed his fleet in the Battle of the Nile on 1 August 1798. In March 1799 the Directory, chronically short of cash and hoping to raise finances by looting the treasures of its neighbours, launched attacks in Germany and Italy. This enabled Britain to form a new anti-French alliance known as the Second Coalition, with Austria, Russia, Naples, Sardinia and Turkey. A Russian Army under Suvorov joined the Austrians in northern

Italy and rapidly pushed the French out in a series of crushing defeats, leaving them holding only the coastal strip around Genoa. Inter-allied political rivalry then caused Suvorov and his army to be transferred into Switzerland, where a second Russian force under Korsakov was operating. Unfortunately for the allies, General André Massena defeated Korsakov at the Second Battle of Zurich on 25/26 September before Suvorov could join him and Suvorov was forced to run for his life up into Austria.

In the meantime, an Anglo-Russian force had invaded the Batavian Republic in August, but had so mismanaged affairs that by 18 October it had to evacuate the place again. Word of these events was allowed to reach Napoleon in Egypt by Captain Sir Sidney Smith, cruising in the eastern Mediterranean with a squadron of Royal Navy ships. The news also included details of the increasing difficulties faced by the ruling Directory in France and its unpopularity. By now, Napoleon was tired of being bottled up in the dust and heat of Egypt; there was no chance of emulating Alexander the Great and invading India and no more glory to be had in Cairo. Sensing that he might well be able to squeeze some personal profit from the political turmoil bubbling up in France, he slipped away from Egypt on 23 August 1799 with a small group of his friends in two frigates and abandoned his army to its fate. He handed command over to General Jean-Baptiste Kléber, who became aware of the honour in a letter which reached him after Napoleon had sailed. Among those abandoned in Egypt was General Louis Desaix, friend and ardent supporter of Napoleon. He was to play a key role in the battle of Marengo.

Luck was with Napoleon and he reached France without being detected by the Royal Navy; landing at Fréjus on 8 October, avoiding the quarantine regulations and reaching Paris on the 16th of the month. After much horse-trading with his old political cronies, Napoleon mounted the coup of 18 Brumaire (9 November 1799) and came, luckily, to supreme power in France as 'First Consul'. He rapidly elbowed the surprised Second and Third Consuls aside and became effective dictator of France.

Having grasped power, Napoleon was as keen as the defunct Directory to fill the coffers and to placate the populace. He decided to mount twin offensives in southern Germany and northern Italy in the spring of 1800. According to Jomini in his *Life of Napoleon*, the Constitution of the Year VIII forbade the First Consul from directly commanding an army. This has recently been shown to be untrue, though Bonaparte was cunning enough to give titular command of the new Armée de la Reserve, which was being formed around Dijon, to the trusty, immensely competent, General Alexandre Berthier, his chief of staff of 1796. For himself, in 1800, Napoleon assumed command of the Consular Guard. If the war in Italy failed, Berthier would take the blame; were it to succeed, why, Napoleon would soak up the glory. Command of the French Army in southern Germany went to General Jean-Victor Moreau, one of the ablest French generals and one-time assistant to Napoleon in the coup of Brumaire. Napoleon kept the war in northern Italy, the scene of his triumphs of 1796, for himself.

Unfortunately, the condition of his 'new' army was extremely poor; the men were short of pay, food, clothing, equipment, accommodation, transport and ammunition. Corruption among procurement officials was rife, attempting to improve matters a Herculean task. Even so, by 22 March 1800, French plans were agreed, although Moreau proved to be very difficult for Napoleon to control, refusing to share command of his army with Napoleon, refusing to advance into Switzerland and insisting on operating in the Danube valley with his 120,000 men against the 95,000 Austrians under Feldzugmeister Baron von Kray. He had also grudgingly agreed to send Napoleon 25,000 men of his army once he had broken the back of the Austrian resistance in Germany. Moreau's offensive also got off to a later start than Napoleon wished; his first actions took place only on 1 May.

In northern Italy, the Austrian commander, 71-year-old General der Kavallerie Michael Melas, was, of course, informed of some aspects of the new French Army by his own intelligence network. It was clear to him that he could expect interference with his operations from this army as soon as the snows cleared the Alpine passes; the only open question was which pass – or passes – the enemy would use. With his 100,000 men, he enjoyed a three-to-one superiority over his foes in the theatre. He planned to exploit this and to attack the French forces clinging on to the coastal strip as soon as the snows cleared from the Appenine passes. This event normally preceded the clearing of the passes of the Alps by about three weeks, so Melas could be reasonably sure that he would suffer no interference with his northern flank during his operations to the south. The Royal Navy commanded the waters and the coast road on the Riviera, and, although co-operation between it and the Austrians was less than perfect, this flank was secure.

On 4 April Melas attacked, cut the French forces into two parts, and by 19 April had Massena and his 12,000 men bottled up in Genoa, while another part of the Armée d'Italie under Suchet was pushed back west over the Var River. So far, so good for the Austrians.

Meanwhile Napoleon concentrated his Armée de la Reserve to the south-east of Lake Geneva. He had a range of passes to choose from for his advance into Italy. He selected the most direct route between his army and its target – the Great St. Bernard – for the main body. Chabran's division would use the wider and better Little St. Bernard, and Turreau's division of the Armée d'Italie would cross the Mont Cenis pass on 22 May to march on Turin.

With his customary energy and genius, the First Consul was organising everything, everywhere from his office in Paris. His plans included intelligence gathering, control and manipulation of the French press, and a campaign of disinformation to confuse his enemies, be they in Britain, Germany, Italy, Austria or France. He sent encouraging letters to Massena, urging him to hold out in Genoa so as to pin the Austrians down around the city for as long as possible. Massena did as he was bidden, though he would finally be forced to surrender on 4 June due to lack of food.

Meanwhile, on 11 April, the Consular Guard, the last formation to move to the scene of the coming *Blitzkrieg* in Italy, left Paris for Switzerland. Bonaparte followed on 6 May. He had just received news of Moreau's victory over Kray at Stockach on 3 May; further victories would follow at Mösskirch two days later and at Biberach on 9 May, as the Austrians were forced back eastwards. Napoleon's back was safe.

On 15 May Lannes's advanced guard of the Armée de la

Reserve started up the Great St. Bernard pass. Unfortunately for Napoleon, his intelligence as to the state of the passes had been too optimistic; this pass was merely a track, impassable to wheeled vehicles or cannon. Much work would have to be done to make it safe and usable as an army supply line but even so the Consular Guard climbed the pass on 19 May and Napoleon followed next day, looking very undignified on the back of a mule.

More frustration for Napoleon quickly followed; both the Little and the Great St. Bernard passes drop into the valley of the Dora Baltea River and join up at Aosta. Some 30 miles further down the river, the narrow valley was almost completely blocked by Fort Bard, perched on an inaccessible rocky outcrop. Napoleon and his staff had ignored reports that this place was a major show-stopper as Berthier wrote on 20 May: 'The fort of Bard is a greater obstacle than we had foreseen.' This fort was held by Captain Bernkopf and two companies of Infantry Regiment Franz Kinsky Nr 47, some 200 men. From 19 May to 2 June, Bernkopf and his men were to block the path of the French artillery. Lannes and the advanced guard infantry and cavalry bypassed the fort by climbing the surrounding mountains, but only six artillery pieces were smuggled through on the road past the fort at night. Napoleon's plan to 'descend like a thunderbolt' had hit a snag.

The next – and final – bottleneck to be carried before the French could burst out onto the fertile plains of the Po, was the defile at Ivrea on the Chuisella River; on 23 May Lannes assaulted the place and took it after a hard fight. Feldmarschall-Leutnant Haddick, defending Ivrea, at once reported to Melas that he was facing 20,000 enemy troops. The uncertainty in Austrian headquarters as to the direction of the French main thrust was now over; Melas, at his headquarters in Turin, knew that Napoleon was descending on his rear. On 31 May, he ordered the abandonment of the thrust against the Var; it was already in difficulties. Four days later, he had concentrated about 28,000 men around Turin to block the obvious thrust to relieve Genoa.

This suited Napoleon well enough. He had just decided to throw Massena and his brave garrison to the wolves and to march east to take Milan, a soft, rich fruit, ready to be picked. This was a hard-nosed decision but undoubtedly the correct one; the First Consul was not going to emulate the Austrian errors of 1796, when they squandered army after army trying to save the fortress and garrison of Mantua. He left Lannes and his force to mask Turin and to create the impression that the relief of Genoa was still his real target. All went very well, and on 2 June the French entered Milan unopposed, Napoleon indulging in childish theatrics in a failed attempt to impress the sullen, silent natives, still hurting from his visitation in 1796. His bulletin reported adoring, cheering multitudes. Milan and its area provided money, food, horses, transport and shelter for his men. It even yielded up 'between three and four hundred cannon – some siege and some field – complete with their carriages.' Some of these were incorporated into the army.

More vitally, Napoleon ordered the switch of his lines of communication from the primitive St. Bernard to the much better developed Simplon and St. Gotthard passes. He was now firmly across Melas's own lines of communication back into Austria, and had seized the initiative in no uncertain manner. He had also split Vukassovich's division from Melas. This division was to be pushed back east to the Oglio River in the next few days.

On 5 June General Moncey arrived in Milan from Germany via the St. Gotthard pass with the reinforcements from Moreau. Unfortunately for Napoleon, there were only 11,500 men instead of the promised 25,000.

In contradiction of his famous dictum 'march divided; concentrate to fight', Napoleon had now only just over 28,000 men under his hand, with a further 22,000 scattered around northern Italy in various detachments. Nothing daunted, Bonaparte leaked information to Melas via the double-agent, François Toli, that he now had 80,000 men instead of the real 50,000. In reality, the total of French troops in Italy at this point was 80,845.[1]

Melas abandoned Turin on 6 June, heading eastwards for the fortress city of Alessandria, between the Tarano and Bormida

---

1. *Österreichischer Militärische Zeitschrift*, 1823, Part III, Heft 8, Page 152.

Rivers, to clear the enemy from his communications. Elsnitz's force from the Var was also ordered to withdraw there. This he did, but not before he had lost quite heavily to Suchet's reinforced army in a series of actions on that river and in the mountains on the way. Hohenzollern commanded the Austrian garrison now established in Genoa and the remaining troops of the old besieging force now split into two parts. Ott's division went north through Novi and then turned east to make for Piacenza; Gottesheim's brigade marched north-east, also to reinforce Piacenza, but O'Reilly was to abandon that place that same day, in the face of French assaults, and to withdraw westwards, towards Melas.

On 9 June, Lannes, with 12,000 men, surprised Ott (11,000) and defeated him at Casteggio-Montebello near the swollen Scrivia River. Ott also fell back to the west on Melas's main body in the fortress of Alessandria.

Despite his forces being badly scattered, Melas decided to offer battle east of Alessandria to the Armée de la Reserve, before it could be joined by the Armée d'Italie, which was coming up from the south-west. The Austrian commander had in fact left garrisons in the following places at the strengths shown: Genoa 5,800; Cuneo 4,390; Turin 3,860; Coni 4,390; Mantua 3,500; Casale 1,560; Cevi 800, Milan 2,816; Tortona 1,200. Small garrisons had also been put into Acqui, Ancona, Asti, Florence, and Piacenza. Melas even left 3,000 troops in the fortress of Alessandria. In these and other places over 46,000 Austrian troops were locked up in Italy's fortresses. He thus had 23,000 infantry and 7,780 cavalry available for the coming battle. Napoleon had 28,170 men in all. Desaix had arrived at Napoleon's headquarters on 11 June from Egypt; he was given command of the divisions of Boudet and Monnier, and the other formations were shuffled about to compensate for this change in the order of battle.

By the evening of 13 June, the opponents were closing up for the decisive battle. But the actions of the last weeks, with the Austrian high command blundering about, wrong-footed as the French ran rings around them, had radically altered the strategic situation and had left the Austrian morale sagging, whilst that of their enemy soared.

The Battle

This battle included two cavalry actions which demonstrate the amazing potential power of the mounted arm. The first was a minor Austrian charge that wrecked the Consular Guard Grenadiers partway through the day; the second was Kellermann's famous deed which decided the battle – after Desaix had saved the day for Napoleon, who would otherwise have managed to throw it away.

The walled city of Alessandria sits on the flat, waterlogged plain of the upper Po valley, protected on three sides by the Rivers Tanaro and Bormida. The citadel of the place lies to the west, across the River Tanaro, linked to the city by a bridge. The village of Marengo lies about three miles to the east of the city, across the Bormida and just to the east of the Fontanone brook. Although not a major waterway, this latter feature was of uneven depth, boggy and overgrown; the eastern bank dominated the western. It was an obstacle, in parts, even to the movement of infantry. The nature of the battleground was patchy; in places it was heavily cultivated, frequently dissected by ditches, hedges and walls, partially covered with mulberry groves and maize, which was already high enough to obscure the sight of a dismounted man. The terrain was scattered with solidly-built, easily defendable villages and, having few roads, favoured the use of small infantry formations and minimised the advantages of the superior Austrian cavalry over large areas. Other parts were flat, firm and open; ideal for cavalry. Control of any battle here would be extremely difficult; much would depend on the initiative of junior commanders.

Just before 14 June Melas heard that French forces under Chabran were also on the north bank of the Po; the feared encirclement preyed on his mind. Together with his chief of staff, Zach, Melas aimed to offer battle to clear his way through Napoleon's army and on, back to Milan and Mantua. Zach was also ordered to leak false intelligence to the enemy that the Austrians planned to cross to the north bank of the Po and that Hohenzollern was to join them from Genoa. To support this deception, a bridge would be built at Casale (north of Alessandria and on the Po) and boats collected at Valenza, north-east of Alessandria on the same river.

Melas's first target after winning the battle was to relieve Tortona on the Scrivia. To this end, he sent O'Reilly's light division off on the evening of 13 June, over the Bormida and along the road south-east, towards San Giuliano. The Austrian army was camped between the city and the Bormida, with a fortified bridgehead across that river. Morale rose as warning of the coming battle was given.

Napoleon received the false information of Austrian intentions late on 13 June. His patrols had reported the phoney bridging activity, and it seems that the First Consul accepted the tale as being true. He alerted his flanking formations to prepare to counter the enemy moves, and detached some of his troops to reinforce these formations. This action reduced the forces that he would have for the unforeseen battle of 14 June. Napoleon was about to be surprised.

On the evening of 13 June Napoleon's army was to the east of the Bormida, scattered from Marengo back to the Scrivia River at Torre di Garofoli. To pin down the Austrians for the next day, Napoleon sent Victor to attack the bridgehead on the Bormida. Victor's troops clashed with O'Reilly's men in Cassina Bianca. To avoid being cut off by the superior enemy force, O'Reilly withdrew to the bridgehead. Gardanne followed up as far as the Pietrabuona farm, but was halted there by Austrian reinforcements and the artillery in the bridgehead. Despite this check, the Fontanone brook in front of the village of Marengo – through which the intended Austrian main axis of advance ran – was now in French hands.

That night, Austrian staff officers saw the campfires of Lannes's men south of Castel Ceriolo, but not those of Victor's further to the south. They thus assumed that the French Army was north of its actual position. Due to effective Austrian cavalry screening, Napoleon was still not certain as to Melas's location or intentions but he suspected that the Austrians might try to slip away without a fight. To close the possible escape route to Genoa, he sent Desaix, who was at Torre di Garofoli, off south with Boudet's division to Rivalta and Novi, even though French cavalry patrols had already reported that area to be clear of enemy activity. Lapoype's division was ordered to move to join Chabran's north of the Po, to close off the road to Milan.

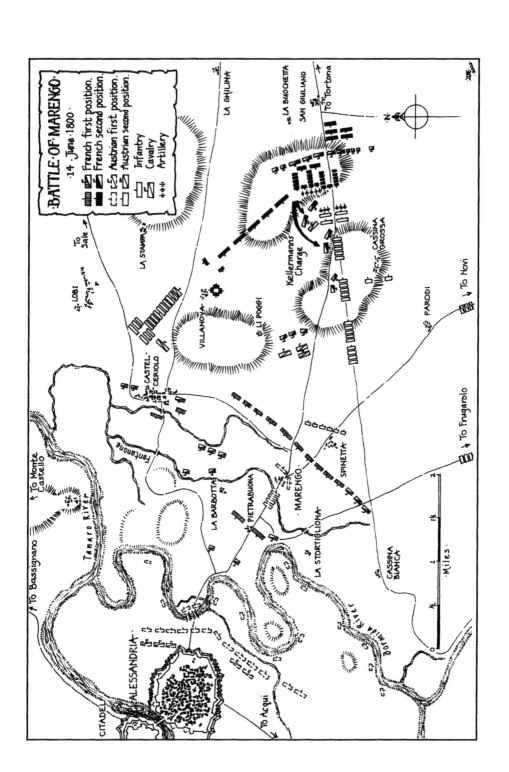

There were two bridges linking the Austrian bridgehead over the Bormida with their main forces and there had been a pontoon bridge further north, which Ott was to have used. On the evening of 13 June, Marmont (commander of the Army of the Reserve's aertillery) had discovered this structure and shelled it. His attack was beaten off by the fire of the heavier Austrian guns, but Napoleon, hearing of the bridge, gave orders for it to be destroyed. One of the Austrian commanders ordered that bridge be moved upstream into the bridgehead to protect it. This meant that, on the morning of 14 June, all Melas's troops would have to use the bridgehead to cross the Bormida, and there was only one narrow gate through which they could pass forward from the bridgehead into battle. The deployment of the army would thus be excruciatingly slow. Some sources state that the Austrian advance out of the bridgehead did not start until 8 o'clock; Arnold, however, says that, 'At daybreak, the main column cautiously filed through the bridgehead's only gate.' This would be normal practice. Melas's plan was for a frontal assault to break the French line at Marengo, with a left hook to turn their northern flank.

According to Esposito and Elting, Austrian orders for the assault were so ill-defined that no formation was given the specific task of taking the key village of Marengo, with its bridge over the Fontanone brook. Be that as it may, the tardy, slow emergence of the Austrians through the bottleneck, gave the French plenty of time to mount counter-measures. Berthier was surprised by what he saw pouring out of the bridgehead; this was no retreating enemy. He left Kellermann to hold the line and rode back to Napoleon – who was in the rear at San Giuliano – to report on the surprisingly serious turn of events. So convinced had the two French commanders been that Melas was going to run, that Napoleon, at about 9 o'clock, sent Lapoype's division of 3,500 men off north across the Po. Berthier, too, had been sloppy in his dispositions and did not secure his flanks. The southern did not extend to include La Stortigliona; the northern ended at the village of La Barbotta.

As an opening move, Frimont's men retook Pietrabuona farm from Gardanne's men and pushed on against Chambarlhac's division. Not until O'Reilly's column began to repeat its drive to the south, along the Bormida, did Victor think to extend his left, by putting a detachment of the 44th Line and a gun into La Stortigliona under command of Achille Dampierre, one of his ADCs. This detachment was later joined by the 11th Hussars. No-one took action about the open, northern flank, because Ott's left column had not yet appeared on the field.

By 10 o'clock, Victor had been reinforced by Murat's cavalry and Lannes was now at Casa Buzana, about two miles behind Marengo. Opposing strengths at the front were now 30,000 Austrians to 15,000 French. Frimont's advanced guard of the main column led the way out, then deployed to north and south, to allow Haddick's infantry to close up to the Fontanone brook opposite Marengo; Kaim's division came up in support. This section of the Fontanone obstacle formed a concave killing ground into which the Austrian assault was funnelled. The fighting was intense, the courage of the Austrians impressive, but the strength of the defences, and the determination of Victor's men, foiled the offensive.

The 43rd Line held Marengo itself and the village layout made it into an ideal small fortress on its western face. As they closed up to the Fontanone brook, the Austrians were caught in effective crossfire from Marengo and from Gardanne's men to the north of that place, and suffered heavily. Haddick, in the forefront of the fight, was hit and mortally wounded. He died in Alessandria a few days later. The leading Austrians fell back through Kaim's division and Bellegarde took over command of the main column. He at once organised a second assault and sent in Kaim's division.

Leading from the front, Bellegarde broke through the Fontanone brook to the north of Marengo and took the farm of La Barbotta. At this point he, too, was hit, and had to be taken to the rear.

Just then, the leading regiments of Lannes's corps reached the farm. The 22nd Line threw the tired Austrians back and occupied La Barbotta; part of the 6th Light made for the northern village of Castel Ceriolo. The 28th and 40th were held as reserve. Watrin's division also mounted a counter attack and forced Bellegarde's column back over the Fontanone.

At about 11 o'clock Napoleon and Berthier returned to the

French line behind Marengo. By this time, the Austrian artillery was beginning to win the upper hand in the centre and Ott's Left Column had managed to squeeze out of the bridgehead and was heading to the north-east, around the open French flank to seize Castel Ceriolo. Frimont's infantry of the advanced guard of the main column joined them. The close terrain hindered deployment, but just before 12 o'clock, General Gottesheim's advanced guard entered the village and threw out the 6th Light.

At one glance, Napoleon realised that, if Ott were to continue unchecked, he would turn the French line. Ordering Monnier's division off to beat Ott's column in the race for Castel Ceriolo, Napoleon sent couriers galloping off to recall Lapoype and Desaix, with the famed 'For God's sake come back' message.[2]

When he arrived at Castel Ceriolo at about 12 o'clock Ott's quick inspection of the terrain to the east proved that it was empty. Feeling that he could not advance further while the Austrian centre was still held up behind the Fontanone, he decided to turn right and attack Lannes's flank. This was the right choice. He did, however, send two squadrons of the 10th Light Dragoons further along the road to Sale to locate the enemy. They later clashed with Rivaud's hussars. With his remaining troops, Ott remained in and about Castel Ceriolo, sending four squadrons of light dragoons to Buzana, deep in the French rear; Gottesheim drove south with Ott's advanced guard.

As the Austrians loomed up from the north, Lannes pulled

---

2. Some sources relate Captain Coignet's apocryphal story that the fighting around Marengo was so fierce, that General Chambarlhac lost his nerve and galloped off to the rear, not to be seen at the front again that day. Coignet also goes on to say that some of the general's own men fired at him next day. This tale is not corroborated by other accounts, but is often repeated. Esposito and Elting quote it, stating that 'Bonaparte showed no mercy to any officer who failed him there. Chambarlhac and Monnier disappeared from the active list.' If this episode did occur, Chambarlhac certainly deserved to be court-martialled and at least cashiered. In fact, according to Six, it did little to stunt his career; he was promoted général de division on 28 July 1803, made a commandant of the Legion of Honour on 14 June 1804 and created a baron of the Empire on 30 June 1811. He died in his bed on 5 February 1826 in Paris; scarcely the career of a proven coward under the unforgiving Napoleon. However, his name is not inscribed on the Arc de Triomphe.

the 28th, 40th and a battalion of the 22nd out of the defences of Marengo to form a new flank. The 28th were almost out of ammunition; as the Austrian cavalry approached over the open ground in this area, they formed square. This square has in the past been mistaken for the Consular Guard, but, according to an eyewitness, Petit, this has now been thrown into doubt.[3] This left only the 43rd still holding the Marengo line.

Luckily for Napoleon, Desaix had been delayed in crossing the Scrivia River by the rising waters. The sounds of the battle to his rear also caused him to drag his heels. Napoleon's desperate cry for help reached him at about midday and he set his troops in motion for the battlefield as fast as they could march.

At about noon, Melas, who was now behind the Austrian front line, saw that his central thrust was making very little headway. Not yet aware of Ott's good progress at Castel Ceriolo, he decided to break the deadlock by turning the southern French flank. He also ordered Major Graf Hardegg to bring up his bridging team to create another crossing over the Fontanone.

Just east of La Stortigliona was a narrow, but unguarded, ford over the Fontanone; Melas ordered Pilatti's cavalry brigade of the main column to force this and to attack Gardanne's southern flank. In fact, this task should have been carried out initially by infantry as the crossing was so restricted. It was so difficult that it took some considerable time for the Austrian cavalry to worm their way across. They were still forming up when they were discovered and promptly charged by Kellermann's six cavalry regiments (his own brigade plus other units). The impact on the surprised and disorganised Austrians was disastrous. They were hurled back into the swampy, overgrown stream, tumbling over one another; many drowned. Weidenfeld's grenadier brigade was rushed to stem the French attack, which seemed to threaten O'Reilly's column on its way south.

Meanwhile, Kaim was organising the third assault on

---

3. Until the publication of Dave Hollins's book, using sources hitherto not cited in English, this unit has been accepted as being the Consular Guard though that version of events was given by Coignet, who was on the southern part of the battlefield near Marengo and could scarcely have seen it himself.

Marengo, with Bellegarde's division to his left. At last, Hardegg's pioneers and pontonniers struggled through the mass of troops to the front. With great dedication, these technical troops formed living bridges, some of them up to their chests in water, for their infantry companions to run over. Why no one had shown such initiative before is a mystery. In a short time, a battalion of Infantry Regiment Nr 63 Erzherzog Joseph scrambled across just north of Marengo.

To their rear, the bridges were built and more men rushed over, including Bellegarde's division which now engaged the 22nd Line. A fire-fight now broke out with the 6th Light and the noise alerted Lannes to the new threat. But, after three hours of combat, Watrin's division was very short of ammunition and getting worn down by the odds. They broke to the rear, Frimont's cavalry in pursuit. Lannes mounted a fierce counter-attack with Rivaud's battalions of Victor's last reserve, the 43rd and 96th. They drove back Infantry Regiment Nr 63 and then poured volleys into the flank of Lattermann's grenadiers. The Austrians were driven out of Marengo and the two sides then stood, firing into each other at close range, just east of the Fontanone. Both Lattermann and Rivaud were wounded in this action.

It was now just before 2 o'clock, when O'Reilly finally threw Dampierre and the 44th out of La Stortigliona and chased them south to Cassina Bianca with Colonel Schustekh's 8th Hussars and the 1st Battalion of the Oguliner Grenz Regiment. O'Reilly attacked Gardanne with his remaining Grenzer regiments.

Austrian pressure against Marengo mounted; at about 2:15, Nobili's cavalry crossed the Fontanone and the 43rd and 96th abandoned the village. Kellermann charged to restore the situation and halted the advance of the enemy cavalry. To the north, Lannes was still holding Gottesheim at bay so Gottesheim rode back to Castel Ceriolo to ask Ott for extra men. He was given Schellenberg's division which set off straight away. But now Gardanne's men were falling back. Victor had sent in his last reserves and ammunition was almost exhausted; he decided to pull his line back to the area of Spinetta and its vineyards. Lannes ordered Watrin's division to pull back in line with Victor. Melas saw the enemy retire and decided to seize his chance. The bridge at Marengo was hastily repaired. Melas led the 1st Light Dragoons Kaiser through the village and the grenadiers followed. To the south meanwhile, O'Reilly had taken Cassina Bianca; the 44th had surrendered. Schustekh's cavalry drove the 11th Hussars eastwards along the road towards Lungafone and San Giuliano; O'Reilly headed off south as he had originally been ordered.

At about 3 o'clock, Napoleon had come forward once again from his headquarters at San Giuliano to see how things were going; he soon recognised that they were going very badly. To the south, Austrian troops were advancing on Spinetta; in the north, Gottesheim was pushing down on La Barbotta and four squadrons of the 10th Light Dragoons Lobkowitz were coming down on Champeaux's men. Monnier's division was advancing on La Barbotta, but the leading troops, the 19th Light under Schilt, fearing the Austrian cavalry there, veered off to the north, into Castel Ceriolo. Here, at about 3 o'clock, they quickly threw out the small garrison, which Ott had left there. Made aware of this, Vogelsang advanced on the place with Ulm's brigade and the outnumbered 19th fell back east on Villanova.

In front, Melas's advance continued. Many of Victor's wounded troops were straggling away to the rear. The situation needed urgent action. Napoleon's limited reserves, including the Consular Guard, were ordered up from San Giuliano. A battalion of the 72nd Line was sent to prop up the northern flank; the grenadiers of the Consular Guard were directed to Victor's right, carrying cartridges to replenish the supplies of the weary troops.

Having completed their task of resupplying the ammunition, the Consular Guard Grenadiers were then sent north to counter Schellenberg's thrust from Castel Ceriolo. They were in open column when they were charged by the 10th Light Dragoons Lobkowitz; with steady volleys they drove the cavalry off.

The following account of the baptism of fire of the Consular Guard is based largely on Dave Hollins's recent research and several hot debates on the internet. It flies in the face of conventional wisdom. No apology is made for including it here.

Champeaux's cavalry, led by Murat, now came up in support of the Guard infantry. As they advanced, they posed a threat to

Sticker's brigade of Schellenberg's division. The leading battalion, the 1st of Infantry Regiment Nr 51 Splenyi, was not impressed. It quickly deployed into line, gave a cheer and bayonet-charged the French cavalry. Getting to close range, it halted and fired a steady volley; Murat and the dragoons fled. The 51st continued in line to attack the Consular Guard Grenadiers, together with Infantry Regiment Nr 28 Fröhlich. A fire-fight developed between the opposing infantry across the Cavo ditch at a range of about 30 yards.

The consolidation of Lannes's formation to the south around La Fournare, had now opened a gap of about 500 yards between them and the Guard. Monnier's division was withdrawing southeast; the Guard was becoming isolated. Oblivious to their peril, the Guardsmen held on. It was at this point that the cavalry of Oberst Frimont's advanced guard (1st Light Dragoons Kaiser and the Jäger-zu-Pferde Bussy) saw their chance and took it. They crashed into the left rear of the surprised line, hacking and slashing to great effect. Infantry Regiment Nr 51 Splenyi joined in the fray. In 15 minutes, some 400 of the Guard were casualties; the rest fled to the rear.

How the several French commanders present – including Napoleon – could have been so remiss as to forget their elite troops and let them get into such an exposed situation is difficult to understand. It seems that the morale of the French troops at this point was so low, that even Napoleon's golden tongue could not move the 72nd Line to go to the rescue of their isolated comrades. Lannes ordered Bessières to charge with the cavalry of the Consular Guard but this move fell on equally stony ground in the face of the muskets of the two battalions of Infantry Regiment Nr 28 Fröhlich.

At about 4:30, by general consent, the First Consul and his depressed troops began to fall off to the east. Seeing this general rearward movement of the enemy, Melas, his age and two falls from his horse beginning to tell, handed over overall command to Feldmarschall-Leutnant Kaim (Melas's chief of staff, Zach, was leading the pursuit of the French) and rode off slowly to Alessandria to write his report.

Just at this black moment, Desaix's ADC, Anne-Jean Savary, rode up to Napoleon with the news that Desaix would soon arrive. He actually came in at 5 o'clock; Boudet's division was 30 minutes behind him. 'Well, what do you think of it?' a nervous Napoleon asked his white knight. 'This battle is completely lost', replied Desaix, 'but there is time to win another!'

A plan was quickly cobbled together. A single battery was formed of the 16 remaining guns and placed in the apex formed by the junction of the old and new roads, north of Cassina Grossa; the three battalions of the 9th Light of Boudet's division advanced in line north of that village to oppose the Austrian advance. Monnier's division, which was to the rear, was pulled back and sent forward on the north flank together with the tired cavalry of Kellermann and Champeaux.

General-Major St. Julien's brigade advanced in line, Infantry Regiment Nr 11 Oliver Wallis leading, Lattermann's grenadiers in line behind them. The triumphant Austrians were getting out of formation, thinking only of finishing off the enemy. Despite being warned of this danger, Zach urged them on. The following action took place just west of San Giuliano, in an area partially planted with new vines, partially flat and open and good for the use of cavalry. The 9th Light conducted a grim, fighting withdrawal to win time. As Infantry Regiment Nr 11 came on in column, it was hit by volleys from the 9th Light, concealed from view among the young vines. Trying to deploy into line, the Austrians fell into confusion and retreated through Lattermann's grenadiers. These were in line and advanced steadily, supported by the regimental guns and by three cavalry batteries, which Zach brought into play. The 9th Light now had to fall back slowly as Desaix rode up to join them.

At about 5:45, Desaix ordered Kellermann to take the remnants of his cavalry from behind the 9th around to the northern flank, to plug the gap between that regiment and Monnier's division, which was still well to the rear. Kellermann moved off with his 400 cavalrymen and came up on the northern flank of Zach's column. Kellermann had the remnants of the 2nd, 20th and 21st Cavalry, 1st, 8th and 9th Dragoons. The Austrian artillery ceased fire at this moment, to allow the 9th Light Dragoons Liechtenstein to charge.

Marmont's 16 guns blasted the advancing Austrians for a last time. This discharge threw the flank battalions of the grenadiers' line into confusion; Kellermann saw his chance and struck, just as the 9th Light advanced again. Hit in front by artillery and infantry, and by cavalry in their disorganised flank, Lattermann's grenadiers were bowled over and broken in a matter of moments. Their formation dissolved completely as panic flared through the ranks. Rushing on, Kellermann hit the two squadrons of the 9th Light Dragoons, who were also in confusion as an ammunition wagon had just blown up in their vicinity as their artillery fell back past them in a mad rush.

As luck would have it, Zach had just ridden up to the Light Dragoons to order them to charge; in the ensuing chaos, he was captured by the 2nd Cavalry. On the Austrian right, St. Julien was also captured, but rescued by Corporal Altmann of Pilatti's 4th Light Dragoons Karaczay. The whole Austrian front fell to pieces in an astounding twist of fortunes as regiment after regiment fled to the rear, shocked, demoralised and leaderless. The elated French surged forward.

Just at this dramatic turning point of the battle, its architect, Desaix, was killed by a shot to the chest. His body was found after the battle stripped of his uniform.

As the defeated Austrians streamed off to the rear, Weidenfeld's grenadiers and Infantry Regiment Nr 63 Erzherzog Joseph were among the last to hold firm in the centre. Pilatti's Light Dragoons were utterly unable to stem the French flood and the disorganised rabble flooded back through the fateful village of Marengo. At about 7 o'clock, one of the last cavalry clashes took place when Murat and Bessières, with the cavalry of the Consular Guard, overthrew the 3rd Light Dragoons Erzherzog Johann.

In the panic to escape, many Austrian artillery teams became stuck trying to ford the Fontanone and the guns were lost. On the southern flank, however, some Austrian units held fast on the San Giuliano–Lungafone road. Infantry Regiment Nr 11 Michael Wallis and the Paar grenadier battalion held off the 9th Light and part of Kellermann's cavalry. To their right, Infantry Regiment Nr 47 Franz Kinsky and two squadrons of the 3rd Light Dragoons Erzherzog Johann under General-Major de Briey at Cassina Grossa had not been involved in the disaster and withdrew in good order on Marengo. O'Reilly also returned

The Battle of Marengo

to Marengo by about 7 o'clock to join Weidenfeld's last stand there. By 10 o'clock these units had withdrawn over the Fontanone and into the bridgehead. On the northern edge of the field, Ott's division, pursued by part of Monnier's division and Rivaud's cavalry brigade, retreated through Castel Ceriolo, crossed the lower Fontanone stream and withdrew into the bridgehead in a relatively intact state.

Austrian losses in this battle, one of the cornerstones of the Napoleonic legend, were 963 killed, 5,518 wounded and 2,921 captured along with 13 guns, 13 ammunition wagons, ten colours and standards. Feldmarschall-Leutnant Haddick had been mortally wounded, Generals Bellegarde, Lattermann, La Marseille and Vogelsang had been wounded. As usual, Napoleon admitted to far fewer casualties than had actually been suffered. Berthier later gave French losses as 1,100 killed, 3,600 wounded and 900 captured or missing. General Desaix had been killed; Champeaux mortally wounded.

Back in Alessandria, the morale of the Austrian high command had been broken by the stunning reversal of their fortune. Melas was in shock. He had no more stomach for another battle, and had only 19,000 men still available. He capitulated the next day, and was allowed to evacuate his army to Austria. Napoleon generously presented him with a sabre that he had brought from Egypt. Melas had committed the 'penny packet sin', spreading too much of his army in garrisons all over the place, thus having insufficient resources available for the critical battle. 'He who defends everything defends nothing.' Melas retired in 1801 and died in 1806.

Northern Italy was again under French control. Thanks entirely to the efforts, skill and courage of his friend Louis-Charles-Antoine Desaix, the fighting spirit of his troops and the timely charge of Kellermann's cavalry, Napoleon had been saved from certain defeat caused by his own poor generalship. Once again, the powerful potential of shock action by cavalry, in the right place, at the right time, had been clearly demonstrated. Kellermann's charge had tipped the scales into complete success after Desaix had stabilised the front. As Marmont later said: 'Kellermann's timing was perfect; had he charged three minutes earlier, or three minutes later, he would have achieved nothing.'

Later that evening, Napoleon thanked Kellermann briefly for his efforts, then turned to Lannes to say: 'I thought the Guard did rather well today.'

Kellermann went on to become one of Napoleon's senior generals and one of the more blatant looters in the imperial pack. Complaints were made to the Emperor about this, but Napoleon would reply: 'Whenever his name is mentioned, I think only of Marengo.'

Kellermann served in the Peninsula until 1811, then was sick during 1812. In 1813 he fought at Lützen, Bautzen, Dresden and Leipzig. On 13 February 1814 he was appointed to command VI Cavalry Corps; he fought at Mormant, Bar-sur-Aube, la Barse and St. Dizier. On 6 May 1814, Louis XVIII appointed him a member of the council of war for the royal guard, and, on 1 June, Inspector-General of Cavalry. Kellermann rallied to Napoleon in 1815 and on 16 March received command of a cavalry division. He was created a peer of France in June and received command of the III Cavalry Corps. He fought at Quatre-Bras and Waterloo. By 3 July he was in Paris with Generals Gérard and Haxo to negotiate with the king in Davout's name, following Napoleon's second abdication. Kellermann was placed on the non-active list on 4 September 1815. On 31 August 1817 he was created Marquis de Valmy and held various senior military and political posts before he retired on 1 July 1831. He died on 2 June 1835. His name is inscribed on the Arc de Triomphe.

As was to happen so frequently in the future, Napoleon felt compelled to tinker with history to put his own spin on it. He altered Berthier's report to show himself in a more positive and dominating role. Not content with this, in subsequent years he altered his account of the battle yet again. He had Desaix uttering patriotic final speeches and gilded the lily mercilessly. Napoleon had been presented with the dramatic, crushing victory that he needed to cement his authority as First Consul but he had neither earned nor deserved it. Even so, on 2 July he entered Paris as the hero of the hour. Marengo was to give him unfettered power in France – and soon in most of mainland Europe.

Two of his generals earned his extreme displeasure this day; Monnier and Duvigneau. Duvigneau had been trapped under his dead horse for much of the night prior to the battle and injured so badly that he could not resume command of his troops. Despite this, he was relieved of his command on 21 June, retired on 15 August 1800 and was re-employed only on 1 January 1814. Monnier was accused of having been too timid throughout the day and Napoleon was so hostile to him, that he was removed from the active list on 13 September 1802, to be readmitted only on 12 June 1814, after Napoleon's first abdication.

By contrast, on 20 July 1800 Napoleon granted Desaix's mother a pension of 3,000 francs for life. Desaix's name is engraved, very fittingly, on the Arc de Triomphe, as, however, is Monnier's.

On 3 December 1800 Moreau was to achieve his own great victory at Hohenlinden over the Austro-Bavarians under Archduke John. In 1803 he was appointed to the Legion of Honour; then was implicated in the Cadoudal plot against Napoleon. In 1804 he was condemned to two years in prison, then emigrated to America. In 1813 he became an advisor to Czar Alexander in that epic struggle to oust his old rival from imperial power and was mortally wounded at the battle of Dresden on 27 August. His name is inscribed on the Arc de Triomphe.

Sources

Arnold, James R., *Marengo and Hohenlinden*, Lexington VA, 1999
Chandler, David, *The Campaigns of Napoleon*, London, 1966
Esposito, V.J., and Elting, J.R., *An Atlas of the Napoleonic Wars*, London, 1999
Hollins, David, *Marengo*, Oxford, 2000
Österreichischer Militärische Zeitschrift, 1822, Heft 10, Vienna
Österreichischer Militärische Zeitschrift, 1823, Heft 9, Vienna
Österreichischer Militärische Zeitschrift, 1823, Heft 8, Vienna
Six, Georges, *Dictionnaire Biographique des Generaux & Amiraux Français de la Revolution et de l'Empire (1792-1814)*, Paris, 1934

Chapter 3
The Battle of Austerlitz
2 December 1805

Many military historians consider this to be the Emperor Napoleon's most impressive battle. This is not just because he won it, and did so very convincingly, but because of the way he used his strategic skill, cunning and the exploitation of his intelligence services to bring the battle about at all. On 1 December 1805 he needed a resounding victory. The spark which lit the fuse to the war in 1805 was the seizure on 14 March 1804 of the émigré Duke d'Enghien by French dragoons, at Napoleon's secret orders, from the Baden town of Ettenheim, which was in neutral territory. At this time there were several plots against the Emperor's life; Britain was paymaster for some of them. The duke had been active against Napoleon, who decided that an example had to be made. The duke was taken to the fortress of Vincennes near Paris, court-martialled and shot on 21 March. A wave of repugnance swept across the ruling houses of Europe. By 11 April 1805 William Pitt, British prime minister, had managed to persuade Austria and Russia to join the Third Coalition against France, though Prussia still dithered on the sidelines.

It was in the period 1803–05 that the Grande Armée was created and the famous system of organising the army into permanent corps was perfected. Prior to 1803 there had been separate French armies in each theatre of the war, under independent commanders, each answering only to the central political government. Co-operation between them was a matter

The Battle of Austerlitz 47

to be sorted out by the commanders concerned. This had much annoyed Napoleon in 1800, when Moreau refused to dance to his tune. Now, in 1805, Napoleon was head of state and supreme commander of the armed forces of France; he would dictate co-operation in future as he saw fit – on land at least.

A French corps in 1805 consisted of a headquarters and staff, two or more divisions of infantry or heavy cavalry, a brigade of light cavalry for scouting duties, and 'corps troops'. The latter would include heavy artillery, transport, engineers, pontonniers, ambulance convoys, and ammunition supply trains. The regiments in this organisation on this campaign were to remain in their parent corps for the next three dynamic, gloriously victorious years.

This does not mean to say that it was Napoleon who invented the corps system. Commanders in several armies had organised their troops in this fashion since the Wars of the French Revolution began back in 1792 and even earlier. The big difference in 1805 was that the French corps were permanent organisations, whereas their predecessors were *ad hoc* formations of regiments, gathered together for a particular action, and shuffled up the next day into other kaleidoscopic groupings. Credit for being the first to employ more permanent groupings should be shared between Napoleon and Moreau, his rival, who seems to have used this system for periods in 1796 and during his successful campaign in southern Germany in 1800. The great advantage of permanent corps organisations is that the units in the framework get to know their commanders and sister regiments. The commanders build up a rapport with the members of their staffs and the formations and regiments they command; they get to know and to trust one another.

At the same time, the staffwork in the Grande Armée was streamlined, standardised and improved. Once again, it would be an inaccurate oversimplification to suggest, as some do, that Napoleon and Berthier, his chief of staff, suddenly invented the modern army staff system on their own. The development of staff work went on in parallel in many armies of the period. The French did, however, speed up the distribution of operational orders, and reduce the workload of the staffs at top level. It was traditional for the staff at the senior headquarters to issue copies of the commander's orders to each and every regiment involved in an operation and to all intermediate headquarters in the chain. By 1805 the French Army had adopted a system whereby the senior headquarters issued orders only to the headquarters of the formations next down the chain of command. These then would repeat the process to their direct subordinates and so on. Thus, the commander's staff would now copy the order maybe seven or eight times as opposed to some 40 or 50 times, and the number of couriers needed to distribute these orders would also be reduced.

The role of Alexandre Berthier, Napoleon's chief of staff, has often been held up as the ideal and the prototype of the modern equivalent function. I find it difficult to agree with this proposal. A modern chief of staff is required to be an active planning aide to his commander. They bounce ideas off one another; a good staff officer will attempt to correct errors or misconceptions made by his commander if he detects them. Having read about and discussed this topic many times, it seems to me that Berthier may once have been a successful commander of independent mind. His service in continuous close contact with, and under the direct domination of that gigantic character, Napoleon, quickly reduced him to a willing cipher. Certainly by 1809 he was no longer an effective army commander. By 1813 he had been reduced to a mindless but still incredibly effective robot, incapable of questioning his master's decisions, even when, as at Leipzig, he knew that glaring omissions had been made.

From 1803 onwards, Napoleon had concentrated the Grande Armée in a series of camps along the coast of the English Channel in preparation for his planned invasion of Britain. These camps were located at Ambleteuse, Boulogne, Bruges, Étaples and St. Omer. They held 150,000 men. A flotilla of gunboats and transports was built, the harbours along the coast were extended to accommodate the vessels, roads were improved, depots and arsenals built and filled with stores, and telegraph systems installed for the rapid passage of information and orders. There was one large obstacle in Napoleon's path, the Royal Navy. The Emperor ceaselessly harangued his admirals to prepare to seize control of the Channel in order to facilitate the invasion.

His admirals tried in vain to explain to Napoleon that it was not possible for him to order their ships to be in a certain place in the required numbers on a chosen date. The effects of wind, tide and weather all exerted great influence on naval operations, and might well delay them for weeks or months. There was also the fact that the French Navy, its command structure having been wrecked by the mischievous meddlings of the revolutionary government, and having suffered the loss of about 75 per cent of its officers who emigrated in order to survive, had been largely cooped up in its harbours since 1793 by the blockade of the Royal Navy. Many of the officers were unqualified for their posts, the crews were too 'democratic' to obey orders and were often only partly trained. The corps of naval gunners had been disbanded as being too elitist. The ships were quietly falling to pieces, the rigging rotting, the officers and men incapable of adjusting the running rigging in order to move the vessel in the desired direction even if they understood what it was that needed to be done. The Emperor would not – or could not – understand all of this. At one point he ordered a flotilla to put to sea in onshore winds and perform certain evolutions. His admirals protested that it would be folly, a disaster. They were overruled; the boats tried to put out, were blown onto the shore, several were capsized, and over a score of men drowned. Napoleon seethed with rage and frustration.

On 30 March 1805 Admiral Villeneuve, commanding the Toulon squadron of the French fleet, set out to try to achieve the Emperor's aim of seizing control of the Channel to allow the invasion to take place. His cruise was to end in his catastrophic defeat off Cape Trafalgar on 21 October of that year, and in his apparent suicide on 22 April 1806 in Rennes. But Villeneuve's best efforts were not fast enough or good enough for Napoleon. On 23 August 1805, despairing of waiting for his fleet to appear, he threw up the plans for the invasion and gave orders to march his army away from the Channel coast to deal with the emergent threat of Austria in the Danube valley.

Austria was part of the Third Coalition against France, and its armies were to be joined by about 100,000 Russian troops, in three contingents, during the campaign. Other allied diversionary raids were to be made in the Baltic, into Hanover, in northern Italy and in Naples. Thanks to his excellent intelligence-gathering network, the Emperor was well informed as to these machinations, and had developed his own robust answer to them. On 2 September, an Austrian army of 72,000 men under Archduke Ferdinand and Field Marshal Baron von Mack invaded Bavaria, advanced as far westwards as the upper Danube and Lake Constance, and sat down to wait for the French reaction. Mack whiled away his time by repeatedly reorganising the regiments under his command, spreading confusion and dismay throughout his entire army, and convincing himself that his enemy was evaporating day by day.

Napoleon's carefully planted disinformation greased his path. The Grande Armée swept towards Mack and Ferdinand in Ulm in eight corps, concealed behind a very effective cavalry screen. The result was the capitulation of Ulm on 15 October and the virtual destruction of Mack's army, before Kutuzov, commanding the closest of the Russian armies of 36,000 men, had even reached Vienna. Archduke Ferdinand escaped to Bohemia, along with some 12 squadrons of cavalry.

The defeat of the Austrian army on the Danube was not the end of the 1805 campaign, however. In Napoleon's mind, that could only come when he had gutted his opponents militarily and dictated humiliating terms of peace which would ensure his supremacy in mainland Europe. He had ambitious plans for Germany and Britain's turn would surely come.

Initially, it seemed that all was going to plan. Vienna was declared an open city and the Austrian government fled to Hungary. On 12 November Marshals Murat and Lannes bluffed their way into control of the Tabor bridge over the Danube by Vienna, an exceptional achievement, saving possibly days of fighting and thousands of lives. Murat was to incur a severe bout of the imperial wrath three days later, however, when he allowed himself to be fooled in turn, by Russian General Winzingerode at Hollabrunn, into accepting a provisional armistice, halting the French advance and allowing the Russian rearguard to escape. Even so, by 15 November, Vienna and most of Austria were in French hands.

The Austrian forces in northern Italy were on the defensive, but withdrawing north-eastwards to close up with Kutuzov in the

The Battle of Austerlitz

area between Vienna and Brünn in Moravia, where he was awaiting the arrival of Buxhowden with 30,000 more Russians. Bennigsen, with 20,000 other Russian troops, was approaching from the north-east. Kutuzov's rearguard, under Prince Bagration, was fighting bitter delaying actions against Murat, Lannes and Soult from Vienna, northwards on Brünn.

Napoleon decided to concentrate against the Austro-Russian forces about Brünn, as they formed the most serious, and the closest remaining hostile threat. But his army had been marching and fighting for the last three months. Their weapons, equipment, clothing, shoes, horses, transport and ammunition supplies had all suffered considerable attrition. Thousands of stragglers needed to be swept up and returned to their regiments. Thousands of reinforcements were needed from France to replace the casualties and the sick. Added to this was the fact that they were now hundreds of miles from France, deep in potentially hostile country and well into the winter season. Finally, if Prussia joined the coalition, this would threaten Napoleon's deep northern flank. But the allies showed no great enthusiasm to capitulate to him. They needed to be convinced, and convinced quickly, or Napoleon might be caught isolated. Over-extended and outnumbered, he might well be defeated.

The picture was not entirely black, however. As usual, the allies had their several points of difference. The Russians were fairly contemptuous of the combat worthiness of the Austrians. The linguistic problems of communication were considerable. The level of training of the average Russian officer was fairly low, and the Gregorian calendar, used by the Austrians, differed from the Julian, used by the Russians, by 12 days setting another obstacle in the way of simple co-operation. In addition to this, although Kutuzov was nominally in command, Czar Alexander I and his whole retinue were with his general's headquarters. Although militarily inexperienced, the czar would meddle fatally in the operational planning, and his many aides would contribute their tithes to the allied confusion and delay. Better one bad general than two good ones. All in all, it is remarkable that the allies achieved as much as they did under these circumstances.

Napoleon reconnoitred the area around him in detail and finally selected Austerlitz as offering the best stage on which to present his drama: 'The Battle of the Three Emperors'. Napoleon's mind seethed and raced with possible schemes, calculations of manpower, their locations, marching speeds, stocks of ammunition, food, diplomatic initiatives, military limitations. He sent for information, analysed the results on his own, soon saw exactly how he would achieve the great goal that he had set himself, and issued orders for redeployment of his assets to bring it about. The tireless Berthier ensured that the data required by his commander were provided promptly and distributed his orders with incomparable speed and efficiency.

Compare this to the confused, repetitive, interminable conferences which took place on the allied side, where each participant had his own axe to grind, regardless of the common aim, which was probably unclear to most of them anyway. The Austrians lusted to avenge the embarrassment of Ulm, and this played straight into Napoleon's hands. They had about 15,600 men present, but most of the infantry were new recruits. Influential in the allied planning process was the Austrian chief of staff, General Weyrother, who had just taken the place of General Schmidt, who had been killed at Dürrenstein on 11 November.

In order to induce the allies to attack him on his chosen ground, Napoleon divided his army, sending off Davout's III Corps temporarily to the south-west. On the day of the battle Davout was to command Friant's division from his own corps, Legrand's from Soult's corps and Bourcier's dragoon division. Soult and Murat were withdrawn west of the Goldbach stream. Napoleon then invited allied representatives into his lines to discuss peace proposals. The Grande Armée was drawn up in a north–south line, parallel to its line of communication back to Vienna. The mass of the army was in the north, across the road to Brünn; the southern flank was refused and apparently (and deliberately) weak.

The allied army rushed to close with the French, but did not get into position until late on the evening of 1 December. Alexander's planning conference and the issuing of the orders for the battle of the next day took place at 1 o'clock on the morning of 2 December. Weyrother took centre stage, spreading out a large map of the area. His plan was over-complicated, with the

The Battle of Austerlitz

army split into seven corps. Kutuzov napped throughout the performance; only General Docturov seemed to pay attention. The plan was to force the weak French right back, then swing north to cut the French off from their lines of communication and destroy them. But the plan involved the weakening of the allied centre, a fatal error. In fact a French émigré, General Langeron, pointed out this hazard during the conference, but his concern was pooh-pooed.

At daybreak, on a foggy and very cold day, the allies were facing the French along the Pratzen heights, which Napoleon had deliberately presented to them. There are doubts as to the exact number of allies on the field this day. Chandler gives 85,400, Esposito and Elting 85,700, Bowden 72,789. All agree that Napoleon had about 73,000, including Davout's corps, which was marching in from the south-west.

The field of the battle of Austerlitz stretches from the villages of Welatitz and Bosenitz, just to the north of the Brünn–Austerlitz road in the north, to the lake of Satschan, about six miles to the south. From west to east it spreads from the Goldbach stream to the town of Austerlitz itself. The ground is slightly hilly but fairly open, dominated by the Pratzen plateau, with a wide swampy region running north-east from the Satschan lake, along the Littawa River, on the eastern base of the plateau towards Austerlitz.

The allied advance through the fog was a confused affair in which the columns destined to deliver the blow in the south became mixed with one another, delaying the start of the attack. In the south the first, second and third allied columns, some 40,000 men, advanced against 10,000 French troops at 8 o'clock. Buxhowden sent Kienmayer's Austrian infantry to take Tellnitz but the assault was beaten off. More attacks followed and after an hour the French withdrew. Kienmayer's cavalry thrust into the vacant space to harry the enemy. Margaron's light cavalry brigade of Soult's IV Corps fell back before them as ordered. The allied southern wing crossed the Goldbach; all seemed to be going swimmingly. But, at this point, Davout's corps deployed onto the southern front, Bourcier's 4th Dragoon Division in the lead. They ripped into the surprised allies, threw them back over the Goldbach and retook Tellnitz.

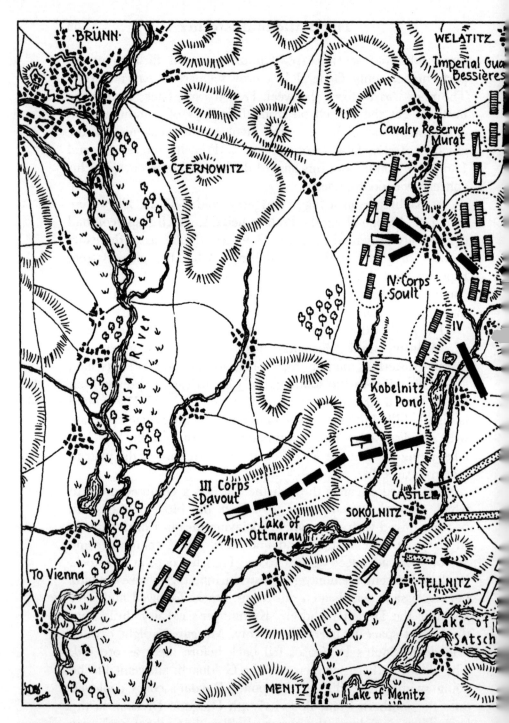

Battle of Austerlitz. This map is after Alison's *Atlas to the History of Europe*. It shows only the first and second acts of the three-part tragedy and does not illustrate the dramatic and deadly finale as the defeated allies scattered before

the victorious Grande Armée, eastwards along the Olmütz road and through Austerlitz and south-eastwards over Lake Satschan

At 9 o'clock Napoleon ordered Soult's IV Corps to push forward onto the Pratzen in the denuded allied centre; Kollowrat's own Austrian division of about 12,000 men from his fourth column was recalled from the southern flank to oppose him at the village of Pratzen. He was thrust back with loss. Soult now swept down into Langeron's right flank. The Russians reeled back to the south; the allied line had been smashed apart. Napoleon ordered a general advance in the centre and north. Bernadotte's I Corps pushed the remnants of the forces of Liechtenstein, Constantine and Kollowrat back towards the allied command post. To the north, Lannes's V Corps and Murat's Reserve Cavalry advanced eastwards against Bagration at about 10 o'clock. Bagration fell back before the onslaught. In the centre, battery after battery of Russian artillery was overrun and taken. The allied army was beginning to fall apart all over the battlefield. Battle was now joined all along the front.

In an effort to stabilise the northern and central sectors of the front – the south was at this point completely separated and in dissolution – Liechtenstein threw the Austro-Russian reserve cavalry into action against Murat's cavalry, 4,000-odd men against about 6,000. Behind the French cavalry, the infantry of Lannes's corps had not had time to form square. The Allied cavalry moved forward alone and without infantry or artillery support. As they closed on Lannes's V Corps, they were met with a volley of canister at point-blank range. This was accompanied by volleys of musketry; shattered, the Austrian cavalry was driven off.

The Russian Grand Duke Constantine Lancers ignored this lesson and charged on alone. As they passed the divisions of Caffarelli and Suchet of Lannes's corps they were raked with volley after volley of close-range musketry. It was magnificent, but it was not war. Some 400 of the regiment were killed or wounded in the space of a few minutes. Lieutenant-General Essen was mortally wounded, the commander of the lancers and 16 of the regiment's officers were captured.

Now Liechtenstein's re-formed regiments, accompanied by Prince Bagration's cavalry, charged again, straight at Murat's command post. They were met by the two carabinier regiments and the 2nd and 3rd Cuirassiers; the crash of the impact was

The Battle of Austerlitz

heard some distance away. The allies were overthrown and driven off.

In desperation, Liechtenstein threw in the remainder of his regiments against Lannes. The French infantry were now in square; they delivered volleys at close range into the allied cavalry. The Austrians were repelled and severely punished for such temerity, as the French cavalry would be at Waterloo ten years later.

The allied cavalry on the northern flank had been thrown into action, unsupported by infantry or artillery, against firm infantry in square, well backed up with artillery and cavalry in a coordinated defence. They had been squandered for nothing.

To the south of Liechtenstein's unfortunate cavalry, the Russian Guard infantry held the village of Blasowitz. Bernadotte's I Corps assaulted them, but was repulsed. Caffarelli's division then attacked the place from the northern side. The 13th Light and the 51st Line threw the Jägers of the Guard out, taking five guns and about 500 prisoners. It was now midday and the allied army was falling back all along the front.

In an excellent example of a co-ordinated attack, Murat's cavalry was supported by Suchet's infantry in a thrust against Bagration's corps. After a bitter fight, the Russians were forced to withdraw, losing some 2,000 men and 16 guns. As they reached the junction in the road to Austerlitz and Olmütz, an Austrian artillery officer, Major Frierenberger, forced his way up through the stragglers with 12 guns. They were rapidly deployed into action on a low hill to the north of the road and halted Suchet's advance for a time.

Bagration's wing was now at Raussnitz, north of Austerlitz. It had lost over 6,000 casualties, but was still a cohesive force. Meanwhile, Soult's men were now attacked by Kaminski's second brigade of Langeron's second column and by the Russian Jägers of the third column. The allies were met with a hail of canister from the French artillery, which ripped great gaps in their ranks. The 10th Light forced the enemy back after a costly struggle. Then Langeron's first brigade came onto the scene, with the Azov Musketeers and the Podolian Musketeers of the third column. They clashed with Thiebault's 2nd Brigade of the 1st Division of Soult's IV Corps, and were then taken in flank by

Legrand's 3rd Division. It was a blood-bath; the Kursk Musketeers lost their regimental guns, their colours and about 1,500 men. Langeron and the Podolian Musketeers were pushed back almost onto Buxhowden's command post at the Sokolnitz pheasantry. Vandamme was now in possession of the Stary Vinohrady hill on the Pratzen, having forced the Austrian Salzburg Regiment (of Kollowrat's division in the fourth column) from it. The fourth column was now a spent force.

Napoleon and his staff advanced to the Stary Vinohrady hill, with Vandamme's division, to control the battle more directly. They arrived to see the Russian Imperial Guard advancing towards them. It was a fantastic sight as the brass mitre caps of the grenadiers glinted in the sun, but the Russians were so keen to close with the enemy, that they broke into a run at 300 yards from their target and were all but spent when they reached the French line. On the way they had been raked with artillery fire and were out of control. Even so, they broke through the first line of Vandamme's men before being repulsed by the second. Despite this reverse, the Russians retired in good order.

Napoleon had sent an order to Vandamme to advance his right wing in order to put more pressure on his foes; Vandamme obeyed but in so doing, he uncovered his left. This chance was seized upon by the Russian command, who at once sent the 17 squadrons of the Guard Cavalry at the target. On this occasion the Russians sent in horse artillery to support the cavalry. The foremost battalion of the 4th Line managed to form square just in time, but the Russian artillery threw canister into it; the square was then hit by the cavalry. The battalion broke and fled, losing its eagle and over 400 men killed. The 24th Light, which was coming up in support of the 4th, formed in line to receive the cavalry and was swept away in its turn. The fugitives swept past Napoleon, but even in their extremity they shouted the obligatory 'Vive l'Empereur!' as they fled.

Now the Russian foot guards returned against Vandamme. Napoleon replied by sending in the cavalry of his Imperial Guard, with a battery of horse artillery to stabilise the situation. In his turn, Grand Duke Constantine sent in the Russian Chevalier Guard and the Cossacks of the Guard. There was a frantic mêlée among the vine poles and casualties were heavy on both sides. The Russian cavalry clashed directly with the Horse Grenadiers of the Imperial Guard but the Chevalier Guard were 'carpet soldiers', more useful at court ceremonial than fighting; they were decimated. Some 200 of the Chevalier Guard, including their commander, Prince Repnin, were captured.

Now, from the north, Drouet d'Erlon's division from I Corps began to arrive to support the French. Four squadrons of Mamelukes and Chasseurs-à-Cheval of the Guard also joined in. They were followed by the Foot Grenadiers of the Guard. Suddenly the Russians broke and fell back on Krzenowitz. If Bernadotte had pushed south on that village at this point, the enemy would have been lost. As so often in his subsequent career, he chose to 'mark time' instead.

It was now 2 o'clock. The allied army had broken into three parts, all defeated, all retreating in divergent directions. With his centre free of the enemy, Napoleon decided to turn Vandamme's division and the Imperial Guard against the remaining foe in the south. Meanwhile, in the south, the Russians were thrown out of Sokolnitz. Overall command from the allied centre had long since ceased, and now command at local level collapsed. The drama on the southern wing of the field had become a tragedy for the Russians as Davout and Soult swept the wreckage of Buxhowden's corps against the frozen Satschan and Menitz ponds and the swamps of the Littawa. General Langeron and some troops escaped to the south over the frozen Menitz pond; the 4th Jägers and the Perm Musketeers tried to hold on in Sokolnitz, but were driven out and scattered, the columns of Kienmayer and Docturov fled over the ice on Lake Satschan. General Przbychevski and 4,000 men were cornered and captured.

Napoleon now ordered his artillery to fire into the ice on Lake Satschan. The stratagem succeeded; the ice broke and a column of Russian artillery of some 38 guns fell through. The skeletons of 130 horses were recovered from the lake when it was drained later in the 19th century.

Casualties were about 9,000 for the French, as well as the eagle of the 4th Line. They lost 14 generals and 594 officers. The allies lost 15,000 killed and wounded and 12,000 captured, as well as probably 180 guns and 45 colours and standards lost. This last figure has recently been called into serious doubt as being badly inflated.

The battle was over; the allies had been decisively crushed and scattered. Napoleon had won the most complete victory of his career in terms of the benefits he would reap through it. His Grande Armée had definitely proved its worth. The Holy Roman Empire would fade into history, and be replaced by the French-controlled Confederation of the Rhine, which would last until late 1813 and give dominance of western Europe to Napoleon. He would redraw the map of the continent to the shapes that we still recognise today.

Sources

Bowden, Scott, *The Glory Years of 1805–1807*, Volume I, *Napoleon and Austerlitz*, Chicago, 1997

Chandler, David, *The Campaigns of Napoleon*, London, 1966

Esposito, V.J., and Elting, J.R., *An Atlas of the Napoleonic Wars*, London, 1999

Martinien, A., *Tableaux pars Corps et pars Batailles des Officiers Tués et Blessés Pendant les Guerres de l'Empire 1805–1815*, Paris, 1890

Pascal, Adrian, *Histoire de l'Armée et de tout les Regiments*, Paris, 1847–58

Six, Georges, *Dictionnaire Biographique des Generaux & Amiraux Français de la Revolution et de l'Empire (1792–1814)*, Paris, 1934

CHAPTER 4
The Battle of Preussisch-Eylau
8 FEBRUARY 1807

Like Gettysburg in the American Civil War, the Battle of Eylau was an unintentional clash, which then flared out of control into a veritable bloodbath. Whereas Gettysburg was a decisive action, Eylau was an exercise in pointless butchery, which solved nothing so that the war went on until the French victory at Friedland on 14 June 1807.

On 14 October 1806 Napoleon had destroyed most of the Prussian Army at the twin battles of Jena and Auerstädt. Actually, the Emperor had tackled the minor part of the Prussian Army under the Prince von Hohenlohe-Ingelfingen at Jena. Some 14 miles to the north, much to Napoleon's surprise, Marshal Davout, with the III Corps, had shattered the enemy main body. Following these hammer blows, the main body of the demoralised Prussian Army fled in utter confusion, to surrender in a series of shameful capitulations over the next few weeks. In a little over a year, Napoleon had defeated the three main European armies: Austria, Russia and now Prussia. He seemed to be invincible.

Napoleon pushed on as far eastwards as he dared without over-extending himself, to seize as much of Prussia as he could and to deny the assets of the territory to the enemy. King Friedrich Wilhelm III of Prussia refused to admit final defeat and fell back eastwards to join up with the Russian Army, which

Alexander had mobilised and ordered westwards to support him. Blücher's Prussian corps of some 15,000 men had survived the catastrophe of 14 October and was now fighting to escape northwards to Lübeck. Bernadotte's I Corps was sent in pursuit. On 6 November, Blücher was cornered near Lübeck. He capitulated next day and Bernadotte was called eastwards to support the Emperor. Napoleon formed a IX Corps to reduce the remaining Prussian fortresses in Silesia and gave the command to his brother, Jérôme.

One other Prussian corps, of about 15,000 men, under General-Leutnant Anton Wilhelm von l'Estocq was still in the field and it marched to join the Russians who were now coming to their aid in eastern Prussia. L'Estocq's corps was placed under command of the Russian General of Cavalry Levin August Theophile Bennigsen. Napoleon now had about 200,000 men in Germany, but, inevitably, many of these were scattered along his lines of communication or involved in operations against the Prussian fortresses in his rear. He needed even more men. Being undisputed head of state as well as commander-in-chief of the army had advantages; he issued a decree to call up, in advance, the conscript class of 1807 from France, and ordered his German and other allies to provide increased contingents.

As well as men, he needed money to finance the war. Germany and Prussia were squeezed mercilessly and provided 560 million francs; British goods were confiscated and either used by the army or sold to raise funds. Cloth, boots, food, fodder, leather, accommodation, horses and transport were all requisitioned in the conquered territory.

The trail of Prussian capitulations continued through November 1806. Those remnants of the Prussian Army still in the field, l'Estocq's corps, were now east of the lower reaches of the Vistula River, at Thorn. By 28 November Bennigsen's army of 32,000 was upriver from the Prussians, between Plock and Warsaw. Off to the east of the River Narew was another Russian army of about 30,000 men under Buxhowden, operating independently. Yet another Russian army of 33,500 men under General Essen was marching up from the distant frontier with Turkey. Napoleon was well informed of all this and aimed to deal with Bennigsen before Essen arrived on the scene.

The Battle of Preussisch-Eylau

Meanwhile, the winter weather turned the entire theatre of the war into a giant, depressingly empty quagmire. The unmetalled roads just dissolved into bottomless stretches of mud. The area of operations was much poorer and less well developed than western Europe and the French Army's traditional system of foraging among the local populace for food yielded only sparse returns.

The first clash of any consequence with the Russians took place at Czarnowo, on the River Bug, 20 miles north-west of Warsaw on 23 December 1806, when Davout's III Corps successfully pierced the extended Russian line along that river in a well-executed night attack, which inflicted a sharp defeat on Ostermann-Tolstoi's 2nd Division. For a brief period the aged Marshal Kamenskoi had assumed command of Bennigsen's army, but he resigned again on 24 December. By three days later Bennigsen had evacuated Warsaw and withdrawn north over the River Bug. L'Estocq's corps formed his western wing at Lautenburg. Napoleon followed up as quickly as the boggy roads would allow, occupied Warsaw and crossed the Rivers Vistula and Ukra on a front of about 50 miles. Ney pushed l'Estocq out of Lautenburg, and turned towards the east, threatening to encircle the Russians.

As was so often to be the case, Napoleon's fragmented enemies had fragmented command structures and divergent lines of communication; the Prussians' lines led north to Königsberg, those of the Russians stretched away to the east.

On 25 December Ney's VI Corps defeated one of l'Estocq's flank guards at Soldau. The Prussians fell back north-east, losing contact with the Russians, who were now in Pultusk. Next day, Lannes and Davout, with 26,000 men, defeated the Russian 2nd and 6th Divisions at Pultusk in a snowstorm. The Russians had 40,600 men present but committed only 29 battalions. French losses were about 4,000; Marshal Lannes and Generals Bonnard, Claparède and Wedel were wounded. The Russians lost about 3,500 men and 12 guns. Both sides claimed a victory, but Bennigsen withdrew northwards, up the Narew.

On 26 December Murat, Augereau and Davout, following up with 16,000 men, caught the Russian rearguard under Prince Galitzin (some 9,000 men of Sacken's division) at Golymin, 12 miles north-west of Pultusk. The Russians were almost caught in a trap, but fought their way out, losing about 780 men. As usual, French commanders were deliberately vague on their losses; 'Equal to those of the Russians' read the report.

The Russians now went into winter quarters; the French did the same. At last, on 11 January, Bennigsen was granted overall command of the Russian troops in the region. The next action took place on 25 January when Bernadotte, with 12,000 French troops, delivered a sharp defeat to Major-General Markov, of Docturov's 7th Division with 9,000 men at Mohrungen, 25 miles south-east of Elbing. During the past weeks, Ney had pushed his corps northwards to the area south of Königsberg, in order to improve the possibilities of foraging. In so doing, he had opened up a gap of some 40 miles in the French line. This was against Napoleon's orders and on 20 January Ney was ordered to return to his original location. Bennigsen also learned of the development and aimed to strike at Ney, hoping to exploit the element of surprise in the depths of winter, but as Ney slipped away south the blow was taken by Bernadotte. A secondary Russian aim was to clear the east bank of the Vistula of the enemy.

Bennigsen now had some 77,000 men under his hand, including l'Estocq's force, and was concentrated around Heilsberg, south of Königsberg, ready to sweep south against Napoleon. Unfortunately, the Russian commander committed the usual sin of sending part of his force, l'Estocq's corps, on a divergent course off to the west, towards Graudenz on the Vistula, one of the Prussian fortresses still holding out. On 23 January the Russians bumped into Ney's surprised cavalry patrols and the secret was out. As soon as Napoleon learned of this he planned a counter-stroke, and ordered a general advance north on Allenstein. By 27 January, l'Estocq had forced General Rouyer to lift the blockade of Graudenz, but the value of this was miniscule compared with that of destroying the enemy's main field army. Once again, the allies had no clear vision of their actual target.

On 28 January Napoleon sent orders to Bernadotte detailing his plan of action. The courier was a newly-arrived officer with no local knowledge of the area; he was captured and the letter

The Battle of Preussisch-Eylau

sent to Bennigsen. In fact, seven copies of these orders were eventually captured by the Russians and none got through, thus demonstrating the value of effective Cossack patrolling. As he read the order, Bennigsen's blood ran cold. He suddenly realised that he was walking into a trap! The Grande Armée was concentrating to strike against his rear. He at once ordered an about turn of his forces and a concentration on Ionkovo. L'Estocq's corps was still at Osterode, some 18 miles to the south-west of the Russians, marching back from its wild goose chase. But now, Bernadotte was in the dark as to the Emperor's plans, and was at Strassburg on the River Drevenz, about 60 miles to the south-west of the impending action. He was to receive no orders until 3 February and would thus miss the bloody battle of Preussisch-Eylau.

Napoleon's orders for the coming battle he planned at Allenstein seriously underestimated the difficulties of moving on the poor roads and the plan consequently failed. On 3 February a clash at Bergfriede on the frozen River Aller gave the French a victory when Leval's 2nd Division of Soult's IV Corps took six guns from Kamenskoi's 14th Division and inflicted 1,100 casualties on it in return for 'minor losses', but this was a sprat rather than the mackerel which Napoleon was seeking. During the night, the Russian Army escaped north towards Königsberg; l'Estocq's corps trailing off to the western side, though as the Russians neared Königsberg they abandoned their lines of communication to the east and risked being penned up there. On 6 February Murat caught up with Barclay de Tolly and his rearguard at Hoff and won a narrow victory, inflicting 2,000 casualties and taking five guns and two colours for the loss of about '1,500' men. The stage was now set for the main battle.

Uncharacteristically for Napoleon, he had failed to concentrate all his available forces for the impending struggle. The corps of Ney and Davout were to arrive only part-way through the action, and that of Bernadotte had no chance of participating at all. Much of what really happened in this battle is unclear, and the existing evidence is often contradictory. What follows is as clear a reconstruction as possible.

As is often the case with famous sites, the field of Eylau is remarkably unimpressive in appearance. It is slightly rolling country, dotted with lakes, ponds and marshes, often connected with small streams. On 7 February 1807, as the battle began, all these water obstacles were frozen hard and most of them were invisible under a thick blanket of snow.

Bennigsen passed through Preussisch-Eylau, making no attempt to hold the place, as he considered it to be too dominated by the heights to the south-west and too vulnerable to envelopment. He drew up his army north-east of the small town, on the forward slopes of the low hills, in deep formations, which presented the French artillery with wonderful targets from across the gentle valley. He was to repeat this tactical madness at Borodino in 1812. Both his wings were hanging in the air for lack of any natural anchor points. Barclay de Tolly's brigade was charged with the temporary defence of Eylau that night. The French Army, in the form of the leading elements of the corps of Murat and Soult, began to appear on the field at about 2 o'clock. Augereau and the Imperial Guard had joined them by nightfall. Marshal Augereau was sick and asked to be relieved of his command, but the Emperor persuaded him to stay on. At this point Napoleon had some 45,000 men and 200 guns on the field; Bennigsen had about 62,000 men and 460 guns. Due to falling snow, this imbalance was invisible to all concerned.

It seems that neither side was planning to open fire that evening, when, at about 3 o'clock in the afternoon, the Imperial equipage blundered into Eylau, apparently unaware that Napoleon had decided to set up camp at the Ziegelhof (brickworks), about a mile south of the town. A Russian patrol chanced upon them and started to plunder the vehicles; they were driven off by the escort, and Soult's troops rushed up in support. Gradually the combat escalated as both sides fed more men in, willy-nilly, to the fighting in the deepening gloom. It went on until about 6 o'clock, when Bennigsen ordered his forces to fall back. Each side had lost about 4,000 casualties.

Soult had fed in the 18th and 46th Line Regiments and they were taken in flank, when disordered, by the St. Petersburg Dragoons. There was furious hand-to-hand fighting, during which, Troopers Deriaguine and Podvoronti of the dragoons took the eagle of the 2nd Battalion, 18th Line. The 18th suffered very heavy casualties, and the French fell back. Bennigsen then

The Battle of Preussisch-Eylau

ordered Eylau evacuated and the French took possession unopposed.

The stage was finally set for the slaughter of the next day. Napoleon's plan was to turn the Russian left flank with Davout's corps (which was coming up on that flank) and to drive them west, cutting them off both from their lines of communication and from Königsberg. Bennigsen's plan was constrained by the fact that l'Estocq was still away to the west. If he fell back east, towards his supply lines, he risked the destruction of the last Prussian field force in isolation.

As 8 February dawned, heavy snowstorms blanketed the area. The exact numbers involved on either side are not definitely known. Estimates of French strength vary from 63,000 at the start to 90,000 after Ney and Davout arrived; Bennigsen is usually credited with 67,000 at the beginning and 75,000 after the arrival of l'Estocq's corps. However, the 19th century German historian Höpfner tells us that l'Estocq's corps reached the battlefield on 8 February with only 5,584 combattants, and that Bennigsen then had only 63,500 effectives, having started the day with 58,000. Both Ney and l'Estocq would appear on the western flank of the field later in the day and in fact they also clashed at Schlautienen on 8 February as they jockeyed to arrive first at the battle.

From west to east, the first rank of the French battle line ran as follows: the divisions of Lasalle, Leval, Legrand, Augereau's corps, St. Hilaire, Milhaud. In second line were the Guard infantry (in reserve), d'Hautpoul, Grouchy and the Guard cavalry. Drawn up in opposition (in the same sequence) were Markov's brigade, then the divisions of, Tutchkov, Essen, Sacken, and Ostermann-Tolstoi in the front rank, with one brigade of Tolstoi's division on the far left flank. Behind them were Somov, Docturov and parts of Kamenskoi's 14th Division. Cossacks swarmed on both flanks. Contrary to popular belief, the Russian commander deployed his artillery in various large batteries. There was one of 60 guns on the right wing, one of 70 12-pounders in the centre of the line, and one of 40 such guns before the left flank.

As day broke, the 60-gun Russian battery on the west wing opened up; Legrand's artillery replied. This surprised Napoleon

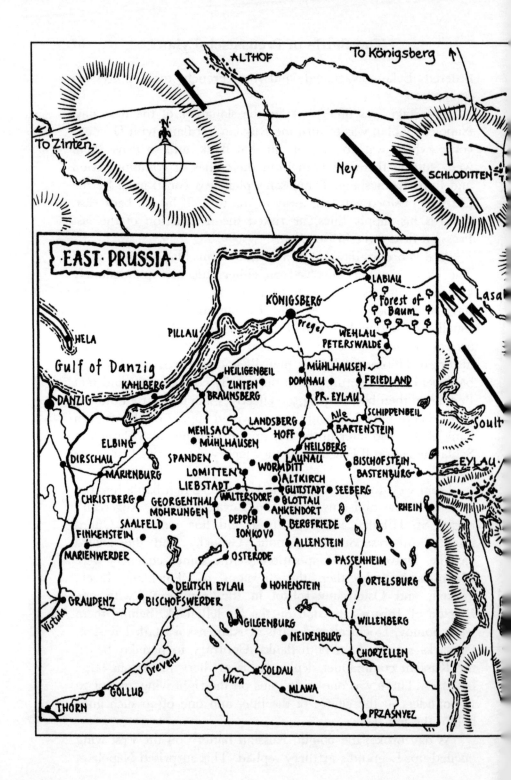

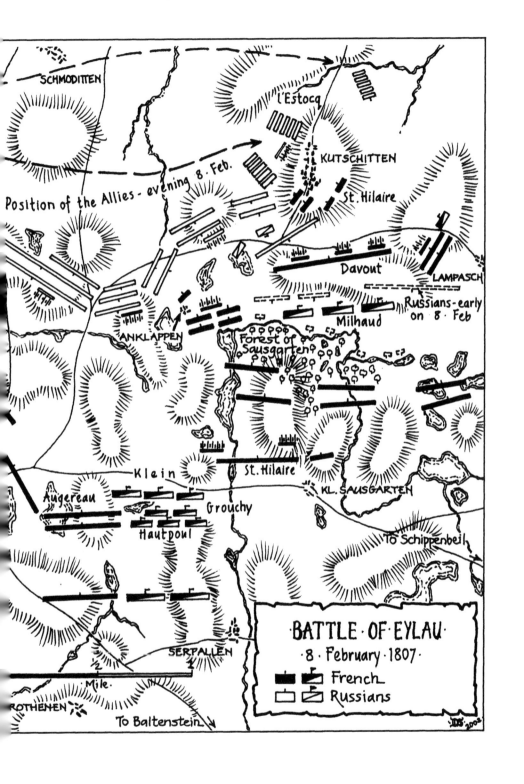

who thought that Bennigsen would fall back again north behind the River Pregel. The artillery duel quickly spread all along the line, but only the left wing of the French line (Soult's two divisions) suffered real harm, as the rest took advantage of the cover offered by the low hills. Legrand's division made a sally in the centre from Eylau and was repulsed.

Delayed by a heavy snowstorm, another French attack was launched at about 9:30 by Augereau's VII Corps and St. Hilaire's division, to support Davout's flank turning movement. The wind came out of the east and blinded the French as they advanced. Augereau lost his way, veered off to the left and lost touch with St. Hilaire's division. When suddenly the snow cleared, Augereau's men were staring into the muzzles of the 70-gun battery in the centre of the Russian line. Canister ripped through the VII Corps; the death toll was dreadful. The plight of the French columns was made even worse by the fact that they were now taking fire from their own artillery as well. When General Essen and his column from the Russian 8th Division delivered a bayonet charge, the men of the Moscow Grenadiers, the Schlusselburg Musketeers and Somov's 4th Division locked themselves into a grim struggle with the equally determined French.

Denis Davidov, the Russian Cossack commander of 1812 and 1813 fame, was present. He described it as like nothing else he had seen, before or since:

> 'For about half an hour you could not hear a cannon or musket shot, only the indescribable roar of thousands of brave soldiers as they cut one another to pieces in hand-to-hand combat.'

Augereau's men fought as hard as they could, but then buckled and melted away, back to their own lines. Prince Galitzin's cavalry harried them along. The Russians broke into the French line at the churchyard near the town, but were thrown out again by Bruyères's cavalry brigade of Lasalle's division and the duty squadron of the cavalry of the Imperial Guard. Napoleon was using this point as a lookout post at this time and was lucky to escape capture.

St. Hilaire's column, meanwhile, had pushed on against

The Battle of Preussisch-Eylau

Serpallen but was repulsed with loss by General Kachovski with the Lithuanian Cavalry and the Little Russia Cuirassier Regiments. The 55th Line suffered particularly badly here and lost an eagle.

All in all the French assaults had failed badly. To restore the situation, Napoleon had only his beloved Imperial Guard, or Murat's Reserve Cavalry. He ordered Murat and 80 squadrons of 10,700 cavalrymen into action; it was now 10:30. Murat moved out in column of divisions: Bruyères, d'Hautpoul, Grouchy, Klein and Milhaud. This was one of the most massive cavalry charges in history. They had to cover about 2,500 yards in driving snow, over ground already covered with snow, which hid many obstacles; it was not done at a gallop.

It was at this point that Colonel Lepic, of the Grenadiers-à-Cheval of the Imperial Guard shouted the famous remark to his men, some of whom were flinching as Russian artillery projectiles passed through their ranks: 'Heads up, by God! Those are bullets, not turds!'

Some sources state that Colonel-Major Dahlmann with the Chasseurs-à-Cheval of the Guard led the way. Höpfner gives that honour to Grouchy's division, which then attacked the Russian cavalry which was cutting down St. Hilaire's division. The massive column scattered Galitzin's troopers as they fell back before them.

Davidov saw this event as well and commented:

> 'The brilliant Murat, sporting his carnival-style uniform and followed by a large entourage, led the attack with bare sabre and plunged directly into the thick of the battle.'

By now, Murat's great column had divided into two parts, one with St. Hilaire's division, the other clashing with those enemy who were busy finishing off the brave remnants of the 14th Line of Augereau's 1st Division. The eagle of the 1st Battalion, 14th Line, was taken by Corporal Outechine of the Vladimir Musketeers. Klein's dragoon division took up with St. Hilaire's men and stayed with them for the rest of the day.

Meanwhile a massive cavalry mêlée had developed in the centre of the line; the Russians drove Grouchy's men back but d'Hautpoul's cuirassiers took the enemy in flank and drove them off. Murat led the dragoons in person. Bessières, with the Guard cavalry, followed the cuirassiers and broke through the Russian line again. Driving on with incredible force, both columns of cavalry crashed through the centre of the first Russian line, breaking several battalions and scattering some. The mass of horsemen went on through the second line of infantry in a wild, boiling mass, and actually got as far as Bennigsen's reserves. Here they were surrounded by clouds of Cossacks and regular cavalry and another furious mêlée burst out. However, many of the French cavalry were mounted on fine, sturdy horses, just taken from the Prussian Army; the lighter Russian horses could not stand up to them.

The French cavalry gathered in the rear of the enemy and charged back through the 70-gun battery, hacking down the gunners as they rode back to their own lines.

Davidov commented:

> 'Two squadrons of horse-guard grenadiers, which composed the tail of the retreating enemy cavalry, were intercepted by ours and laid down their lives between the church and the second line.'

This claim is disputed by many, though Russian cavalry were certainly soon in hot pursuit of Murat's men. The French batteries on the left of their line were overrun by several squadrons and the gunners cut down. As the horse teams had been scattered to the rear, none of the guns could be taken off by the Russians.

Fortunately for the French, the Russians were in such confusion, recovering from Murat's charge, that no infantry were sent forward to support this successful advance. Casualties among French generals in this massive brawl were numerous; Marshal Augereau was among the wounded. Most of the regiments in Augereau's corps had been reduced to cadres. In his own post-battle report, Augereau gave his corps' losses as '929 killed, 4,271 wounded'. The charge had cost the French cavalry 1,500 casualties in a very short time. In view of the fact that it had saved the centre of the French line and thus the day for Napoleon, it was a small enough price. The charge had also caused the Russians massive casualties, weakened their centre,

The Battle of Preussisch-Eylau 73

thrown Bennigsen off balance and won vital time for Davout's corps to come into play on the eastern flank. Without Davout, Napoleon's battle was lost.

By this time, action along the front had reverted to an artillery duel, which went on until midday. Then Davout's corps appeared east and south of Serpallen and drove into Bennigsen's open left flank, driving it back north as far as Kutschitten by 4 o'clock. The critical outflanking movement was succeeding; the trap was closing. The area to the Russian rear was thick with stragglers and wounded making off to the north as more and more Russian battalions disintegrated under the hammering of Davout's artillery. But, on the west wing of the battle, help was already at hand for the struggling Russians.

L'Estocq's weary, embattled corps had outrun Ney, at some cost, and had appeared on the battlefield through Althof at about 3 o'clock. Russian staff officers directed them to the threatened left flank. The Prussians hurried on, gathering up hundreds of Russian stragglers as they went. In the nick of time, as darkness fell, the Prussians stopped Friant's division and Milhaud's light cavalry in Kutschitten. The bitter fighting continued, however, until Ney's corps finally appeared on the field in the west and threw Tuchkov's men out of Schloditten. Making one final effort, the Russians retook the place and Ney fell back.

The gathering darkness was broken by the flames from the burning villages and ripped by the explosions of ammunition wagons. The entire area was covered with dead and dying men and horses; in places the casualties were heaped in layers. After 14 hours of tremendous combat, the day of bloody Eylau drew to a close.

Losses were severe on both sides; what would happen now? Luckily for Napoleon, Bennigsen's nerve broke first when he read the parade states late that night and learned that he had only about 30,000 combattants still with the colours. Against the sometimes violent counter-arguments of his commanders (General Knorring of the General Staff lost his post for his outspokenness) Bennigsen decided to call it a day; he ordered the army to fall back north on Königsberg. L'Estocq, for his part, considered that the maintenance of the Russian communications to the east was the top allied priority. His chief of staff, Colonel

Scharnhorst, agreed with him. Instead of following Bennigsen, he thus withdrew to Friedland on the River Aller, 26 miles southeast of Königsberg. To Scharnhorst fell the unenviable task of informing the Russian commander of this disobedience. There was an almighty argument, but later on, Bennigsen admitted that his Prussian allies had been right.

There was no French pursuit. They were in no state for more serious combat. Only cavalry patrols followed up the allied retreat. Napoleon was relieved. He was also furious that Bernadotte had again failed to support him, as had been the case at Jena–Auerstädt. This time however, it would seem to be no fault of the Prince of Pontecorvo that kept him out of the action, just fine enemy cavalry work.

The cost of this bloody, pointless battle had been very high. French bulletins admitted 1,900 killed and 5,700 wounded but, as we have seen from Augereau's account, his VII Corps alone might have lost this number of casualties. The corps was broken up shortly after the battle, the regiments given to other formations. Eleven days after the battle, l'Estocq's corps returned to the site and reported burying some 10,000 corpses, over half of which were French. Generals Desjardin, d'Hautpoul, Binot, Bonnet, Corbineau, Dahlmann, Lochet and Varé were killed. Bodart gives French casualties as 23 generals, 924 officers and 21,000 soldiers killed or wounded.

The French also lost seven eagles, those of the 1st Battalion 10th Light, 2nd Battalion 4th Line, 1st Battalion 14th Line, 2nd Battalion 18th Line, 2nd Battalion 24th Line, 1st Battalion 44th Line, and one of the 51st Line. There are rumours that another was taken, probably from the 55th Line. As can be seen, Augereau's corps was well to the fore in this collection. Andolenko lists eagles of the 14th, 24th, 44th, and possibly also of the 105th. Bodart gives the French trophies lost as five eagles and seven colours. Esposito and Elting state that only four eagles were lost, but quote no sources for this. The deeper one digs, the more complex the situation becomes. Bennigsen wrote that he had taken six eagles, but returned only five to St. Petersburg, explaining that the soldier who took the sixth had sold it in Königsberg as being made of gold.

Russian losses were Generals Arseniev, Barclay de Tolly,

The Battle of Preussisch-Eylau

Docturov, Gersdorff, Lieven, Korff, Mitzki, Sukin and Titov wounded; Napoleon claimed 18,000 Russians killed and wounded, 3,000 captured and 24 guns. The imperial bulletin also stated that 16 Russian colours and standards were taken. Russian accounts do not support this claim, which has nonetheless often been repeated in later publications, even though no hard evidence has been produced to support it. Latest enquiries seem to confirm that the claim is false. The Prussians lost about 800 casualties.

Action flickered out after this bloodbath while both sides licked their wounds. There was a clash at Ostrolenka on 16 February, but major operations rested until June and reached their climax at Friedland on the 14th of that month, the anniversary of Marengo.

Sources

Andolenko, General C.R., *Aigles de Napoleon contre les Drapeaux du Tsar*, Paris, 1969

Bodart, Gaston, *Militär-historisches Kriegs-lexikon*, Vienna and Leipzig, 1908

Chandler, David, *The Campaigns of Napoleon*, London, 1966

Esposito, V.J., and Elting, J.R., *An Atlas of the Napoleonic Wars*, London, 1999

Höpfner, Edouard von, *Der Krieg von 1806 und 1807*, Berlin, 1851

Lettow-Vorbeck, Oscar, *Der Krieg von 1806 und 1807*, Berlin, 1892–93

Martinien, A., *Tableaux pars Corps et pars Batailles des Officiers Tués et Blessés Pendant les Guerres de l'Empire 1805–1815*, Paris, 1890

Six, Georges, *Dictionnaire Biographique des Generaux & Amiraux Français de la Revolution et de l'Empire (1792–1814)*, Paris, 1934

Chapter 5
The Battle of Albuera
16 May 1811

In this battle the Polish Lancers of the Legion of the Vistula and the 2nd Hussars demonstrated the terrible effectiveness of cavalry when presented with the open flank of enemy infantry in line. Since his arrival in the Peninsula in 1808, Wellington had been slowly and carefully building up his military potential inside Portugal, with the aim of ultimately invading Spain and driving the French out and back into their own country. It was a slow business. His resources were strictly limited and the British and Portuguese troops that he commanded had to be forged into a robust army before he could think of meaningful offensive action. All the while, he also had to fend off thrusts by the superior French Army and attempt to co-operate with his quixotic Spanish allies.

There were only two feasible invasion routes between Portugal and Spain for an army in those days. In the north there was the road from Guarda to Salamanca, and this was blocked by the fortress of Ciudad Rodrigo; in the south there was the road from Lisbon to Madrid and this was commanded by the French-held fortress of Badajoz on the River Guadiana. Possession of both entries into Spain was essential if any element of surprise was to be achieved.

At last, in early 1811, Wellington was confident that he could try to wrest mastery of both gateways from the enemy. Marshal André Massena had finally been forced to admit defeat before the Lines of Torres Vedras and had fallen back into Spain.

The Battle of Albuera 77

Wellington set his forces into motion. He would command the thrust to break in the northern gate at Ciudad Rodrigo; Marshal Beresford, the British Commander-in-Chief of the Portuguese Army, was delegated to break in the southern one. However, from 22 April to 12 May 1811 Beresford unsuccessfully laid siege to Badajoz with a 20,000-strong Anglo-Portuguese corps.

As predicted by Wellington, the French could not allow these key fortresses to fall without making a determined effort to prevent it. Marshal Soult, commanding the French Army of the South in Extremedura, soon became aware of the threat to Badajoz. He scraped together some 25,000 men from his various garrisons and the siege lines before Cadiz, and made ready to march to relieve the fortress. Of this army, 16,000 men were borrowed from Andalucía. Soult considered this force to be adequate to defeat anything that Wellington could detach to operate against him and, anyway, he dared not denude his area of responsibility any more, or the Spanish guerrillas would swarm in behind him and overwhelm the weakened garrisons. He was soon to find that he had miscalculated the opposition that he would face.

By midnight on 9/10 May, Soult had his force concentrated and marched off to the north, hoping to surprise the allies with his speed. The local Spanish guerrillas reported this event so quickly that Beresford heard of it two days later. There were three possible routes which Soult could use; he chose the shortest and most direct, which led through the village of Albuera. According to Wellington's previously agreed strategy, the Spanish outposts in Soult's path made no effort to stop him but fell back to their appointed rendezvous with the Anglo-Portuguese allies at Albuera. Beresford, meanwhile, had at once informed Wellington of Soult's advance, raised the siege of Badajoz and marched off to join his Spanish allies at the rendezvous, leaving part of his force (Kemmis's brigade of the 4th Division) behind to remove the siege equipment to safety before joining him.

On 15 May the allies were drawn up in and to the south of Albuera, though Soult's hussars and chasseurs drove in a strong detachment of allied cavalry from their post at Santa Marta without much effort, much to Beresford's anger.

The Battle of Albuera

The terrain of the battlefield was totally unremarkable and had no dramatic features, but it was the best defensive position against an enemy coming from the direction of Seville. To the French side it was protected by the insignificant Chicapierna brook, which was then flanked by a low, wooded hill, which hid all enemy movement behind it from allied view. The allies were drawn up behind a low ridge behind the brook, devoid of any trees and having only occasional low shrubs. To their front, across the stream, the French would advance through a wood, which also helped to conceal their movements. The allied northern flank was resting in the village of Albuera, their southern flank being open.

The day was very hot, the sky filling with clouds. Alten's 1st and 2nd Light Battalions of the King's German Legion were in the village itself, with Otway's Portuguese cavalry brigade, the 13th Light Dragoons and two squadrons of Madden's Portuguese cavalry (not under command but happening to be present) to their left flank. Hamilton's Portuguese division was behind Albuera and Collin's Portuguese brigade was also there, making 11 battalions in two lines. In the centre of Beresford's line was William Stewart's 2nd Division, ten battalions of the brigades of Abercrombie, Colborne and Hoghton. Apart from the KGL and the cavalry, all troops were drawn up on the reverse slope of the ridge and invisible to Soult. Behind the 2nd Division was the reserve, Cole's 4th Division, which had arrived from Badajoz at 6:30 that morning, and Carlos de España's three Spanish battalions from Castaños's Army of Galicia. The right wing was formed of Captain General Blake's Spanish Army of the Centre of 12,000 men, the divisions of Ballasteros, Lardizabal and Zayas, with Loy's cavalry brigade. The Spanish infantry was drawn up on the reverse slope in two lines. Their cavalry, and that of Lumley, was on the southern flank. Blake placed himself under Beresford's orders.

These Spanish troops had arrived during the previous night, unbeknown to Soult, who considered that they were still coming up from Almendral, some 20 miles away to the south. When he saw Loy's cavalry out to the south, he thought it was that of Penne Villemur of Castaños's Army of Galicia. It is likely that, had Soult known of the presence of Blake's Army of the Centre,

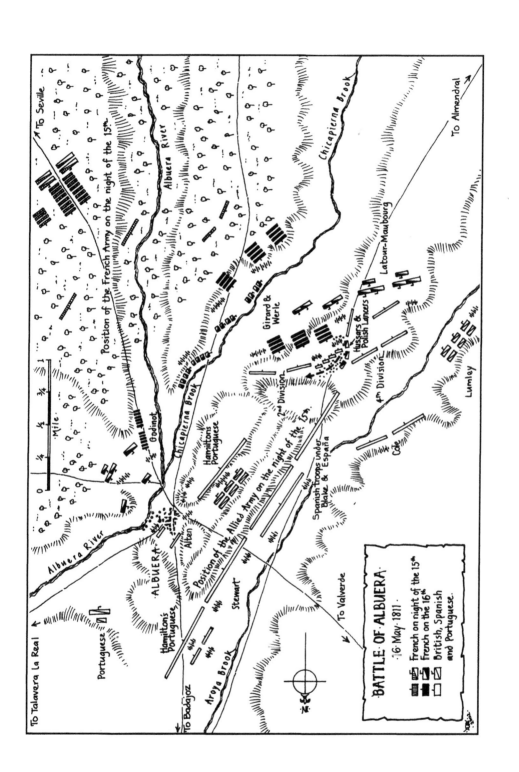

he would not have chanced the action at all. Soult's last brigade, that of Werle, came into position by 8 o'clock on the morning of 16 May. The marshal reviewed the allied position, assuming that Beresford's main force lay between the visible cavalry flank guards. He planned to feint at the village of Albuera to fix Beresford's attention, then to launch his main assault in a wide left hook around the allied southern flank. The northern feint would be carried out by Godinot's infantry brigade and Briche's light cavalry brigade. His main body, V Corps, and the two remaining cavalry brigades under Latour-Maubourg, would be the southern hammer blow, their advance hidden by the woods. Between these two outer groups, Soult posted Werle's 5,600-strong brigade of infantry and the two dragoon brigades of Bron and Bouvier des Eclaz, in full visibility, posing an obvious threat to Beresford's centre.

Early in the action, Colonel Konopka's 1st Vistula Lancers were ordered to make a feint at the enemy line. Four troops of skirmishers from the regiment led the way across the brook, south of Albuera village. The KGL troops in that place did not fire on the enemy cavalry, but the 3rd Dragoon Guards rode out to meet them. The leading squadron of the Dragoon Guards charged the two leading troops of the Vistula Lancers' skirmishers (50 men under Lieutenants Rogoyski and Wojciechowski) and were overthrown. They fled back on their remaining squadrons; after a short reorganisation, all 370 men of the 3rd Dragoon Guards advanced together. The 50 Poles, veterans of many field operations, fell away before them, but, as in so many other instances in the Peninsular War, the British cavalry dissolved into an excited mob, losing all order. At the right point, the Poles rallied and turned on the leading dragoons and struck again. Again the dragoons were put to flight! The ignominious farce was ended when British infantry fire forced the Poles to withdraw. The 3rd Dragoon Guards had lost 20 men; the Poles some 16, mostly to infantry fire. The skirmishers returned to their regiment. They were soon to be in action again.

Soult's plan, meanwhile, had worked perfectly. While Beresford and Blake stood poised against the threats to their front and left flank, a mass of French, Polish and Spanish cavalry suddenly spewed out of the low hills and forest to their right rear.

The Battle of Albuera

Behind them, the V Corps also came on at a smart pace. The allies had been outwitted and were in a very dangerous position. Beresford rode at once to Blake on the threatened right wing and asked him to send half his infantry to form a new line to face the threat. Blake agreed to act, but in fact sent only one brigade (three weak battalions of the Spanish Guard and the Regiment Irlanda of Zayas's division, plus his only battery), as he was still convinced that the real blow would come from his front. This error was to cost the British infantry regiments present very dear indeed. Beresford rode back north to order Stewart's division to be ready to march for the right flank if needed and sent Lumley's cavalry off at once to prop up Loy's Spanish cavalry, which had been pushed back by the enemy.

Over the next 30 minutes, the V Corps deployed into an assault formation, a huge *ordre mixte*, which was probably the idea of Girard, commander of the 1st Division, who was in command of the corps that day. The front line, the 1st Division, was made up of four battalions, each in column of double companies or divisions, ranged one behind the other. To each side of this central column was a battalion in line, and on the outer flanks a battalion in column. Close, too close, behind this mass, was Gazan's 2nd Division. Its four regiments were deployed in one line, each on a frontage of one battalion. Thus, V Corps was arranged on a frontage of about 500 men, about 800 yards across. In places, it was 18 ranks deep.

By this time, Blake could see that Zayas's three battalions, which had occupied the position on the refused right flank, were about to be crushed by the might of V Corps as it lumbered forward. He sent six more battalions off in haste, but the fight began before they arrived. Meanwhile, the two brigades of dragoons in the centre of Soult's line, followed by Werle's 5,600 infantry, galloped off behind V Corps to join Latour-Maubourg's cavalry on the extreme southern wing. There were now some 3,500 enemy horse and an entire infantry corps on the very weak allied flank. Destruction stared Beresford and Blake in the face.

There was a flurry of orders from Beresford as this great threat mushroomed to his right. The 2nd Division marched off to reinforce Zayas; Hamilton's Portuguese moved into the gap thus created, and Cole's 4th Division, Myer's British and

Harvey's Portuguese brigades, marched a mile to the right to take post behind the allied cavalry. They arrived just in time to dissuade Latour-Maubourg from launching his entire cavalry at the outnumbered allied horse. The V Corps, accompanied by three batteries of foot artillery, rolled forward across a gentle valley towards the heavily outnumbered line of Zayas's men, for all the world like a Roman legion in action, the usual cloud of skirmishers to the fore. When they were about 60 yards away from the Spaniards, the French skirmishers fell back, left and right, to the rear of the corps.

Unusually, instead of just ploughing on through the weak and hopelessly outnumbered Spaniards, Girard ordered his men to employ alternate fire and movement, the front two ranks giving a volley, reloading, advancing a little, firing again. This, of course, considerably slowed the French advance, allowing the allies precious time to rush reinforcements down their line to the critical point. It also condemned most soldiers of V Corps to uselessness, just following their comrades of the front ranks and taking casualties from allied fire. Very soon this scene of terrible destruction was shrouded in clouds of smoke from the muskets; the din was frightful, the carnage awful. Much to the credit of Zayas's men, they did not flinch or run. They stood their ground and pugnaciously returned the French fire.

As this brisk fire-fight began Stewart's 2nd Division, Colborne's brigade leading, began to arrive on the scene. Beresford had sent Stewart off with orders to 'support the Spanish'. Stewart was an officer of courage and, perhaps, too much initiative. He decided that the lumbering V Corps was ideally suited to fall victim to a fire-fight with his troops, which were now deploying into a two-deep line, and that he was the man to pull off this coup. He ordered Colborne to advance around and through Zayas's men and to assault the front and flank of Girard's phalanx. It seems that Colborne asked permission from Stewart to put the right wing of the 3rd Foot into column or square to protect his flank, but Stewart refused this eminently sensible request. At a range of 60 paces, the fire of Colborne's brigade ripped deeply into the thick masses before them. At the same time, four guns of Cleeve's KGL battery opened up on this choice target. The slaughter and confusion in the French column was great. The battalions on the left flank could not stand it and fell off to the rear, scattering down the hill. The men of Colborne's brigade smelt imminent victory in the noise, blood and chaos on that Spanish hillside under a sky black with thunderclouds. They pressed forward through the smoke, over the fallen, intent only on getting to grips with the hapless column.

Latour-Maubourg had been watching the discomfort of his comrades in V Corps' unwieldy phalanx with mounting concern. When he saw Colborne's brigade swing forward out of line to present an open flank, he recognised a golden opportunity and ordered the two regiments closest to him to ride forward and strike their enemy down at once. The Lancers of the Vistula Legion and the 2nd Hussars moved off, some 900 strong. The 10th Hussars and the 20th Dragoons followed them. Suddenly two storms struck Colborne's men. Firstly there was a violent cloudburst that drenched everyone, and caused many muskets to misfire; secondly, the two leading enemy cavalry regiments crashed into their open right flank and rode down the rear of their line, hacking and stabbing as they went. The lances of the Poles were particularly effective against the surprised infantry. Unable to fire due to the rain, the startled survivors of the initial shock had to rely on their bayonets. It was almost no contest and they were cut down in droves.

Ensign Edward Price Thomas, 16 years old, was carrying the regimental colour of the 3rd Foot (The Buffs). After his company commander, Captain William Stevens, was wounded, Thomas tried to steady the men. 'Rally on me, men, I will be your pivot!' he shouted. He was challenged to yield his colour. 'Only with my life!' he replied. He was cut down and killed. After the battle, he was buried by a sergeant and a private of his company; they were the only survivors of the 63 men who were present with the company at the start.

The king's colour of the 3rd Foot was carried in the battle by Ensign Charles Walsh. The staff of the colour was smashed by a cannonball. Walsh was then wounded and about to be taken prisoner when Lieutenant Matthew Latham snatched up the colour. He was at once surrounded by the lancers and a furious fight developed for possession of the trophy. One of the riders dealt Latham a smashing sabre blow across the face but he continued to fight until another sabre blow severed his arm. Accounts of the event state that it was his left arm, but a later illustration seems to confirm that it was the right. Dropping his sword, Latham picked up the broken staff and fought on. Finally, he was cut down, trampled on and speared by several lancers. Despite all this, he managed to tear off the fabric of the colour and to hide it in his coat. Latham and Walsh survived the battle, but were taken prisoner by the French. The colour was recovered from the field by Sergeant Gough of the 7th Fusiliers. The lancers took both pikes. Initially, Walsh was credited with saving the relic, but when he escaped from the French, he told the truth of the events. In February 1813 Latham was promoted to captain and presented with a gold medal by his regiment. The Prince Regent paid for him to have cosmetic surgery to his facial wound, and, when he retired in 1820, he received a pension of £170 a year.

The Poles were particularly savage and stabbed the wounded again and again where they lay. Some of the infantry fled to take refuge with Cleeve's KGL artillery battery, which was nearby. In vain. Within a few minutes the lancers had overrun five of the six pieces.

Only the battalion of the 31st Foot, which had not yet deployed from column into line, was able to survive relatively intact. Despite this, it lost 155 men. In a few minutes, Colborne's brigade had ceased to exist. What more convincing proof of the terrible threat that cavalry presented to unprepared infantry could there be? The victorious cavalrymen careered on along the rear of the allied line, attacking Zayas and his staff, the Portuguese staff and then Beresford himself. Luck was with the marshal, however. Being a large and powerful man, he parried the lance levelled at him then struck the lancer from his saddle. One of his staff then shot the lancer dead. This was the swan song of the dramatic charge. The impetus was now spent, the horsemen scattered and losing momentum. They turned for their own lines leaving a trail of death and destruction behind them, driving over 700 prisoners with them. In this move one squadron of the Poles was badly mauled by fire from the Spanish Infantry Regiment of Murcia, which actually took the Poles' standard. Its whereabouts are currently unknown, but there is a suspicion that it may still be in storage in the Museum of the Cortes of Cadiz, in San Felipe's Oratory.

Two squadrons of the British 4th Dragoons were finally sent at the French cavalry as they rounded up their prisoners. They, too, were bested in the mêlée and driven off with both squadron commanders, Captains Phillips and Spedding, being captured. But the confusion they stirred up had allowed many of the British to escape.

The loss of an entire brigade in so short a space of time might have unnerved any force but remarkably the allies continued to fight as before. It was at this point that Stewart's second brigade, that of Hoghton, came into line behind Zayas's gallant Spaniards and opened fire on the enemy cavalry. Unfortunately, they were mixed up with Zayas's men and some of the Spanish infantry were shot by the British in the confusion. To their great credit, even this did not break them. A Spanish officer pointed out the tragedy to Beresford. Beresford's ADC, Arbuthnot, at great personal risk, rode along the front of the 29th Foot calling to them to cease firing and ended the tragic episode. Ballasteros's 3rd Division now also came up into line to the left of Zayas and joined in the fire-fight.

The sacrifice of Colborne's brigade had not been in vain. Their fire, and that of Zayas's brigade, had halted the advance of V Corps, which was now milling about down the slope in front of the allied line. Girard recognised that his own division was a spent force and ordered Gazan's division to assume the lead. In the resulting confusion, the two formations became badly intermixed and disordered and a pause in the murderous battle ensued. Abercrombie's brigade of Stewart's division now arrived and deployed behind Ballasteros's line. Beresford then ordered the two Spanish divisions to fall back behind the British line as Cole's 4th Division also began to come up. For about 30 minutes, Hoghton's brigade and the 2nd Battalion, 31st Foot, the only operational survivors of Colborne's unfortunate brigade, conducted a vicious fire-fight with the mob of the V Corps down the slope from them. It was a contest between 1,900 men in two-deep line and a mass of perhaps 8,000 men, some in ranks 12 deep.

Perhaps the best English account of this bloody interlude is that of Moyle Sherer, an officer of the 1st Battalion, 48th Foot (the Northamptonshire Regiment):

> 'When we arrived near the retiring Spaniards, and formed our line to advance through them towards the enemy, a very noble-looking young Spanish officer rode up to me, and begged me, with a sort of proud anxiety, to take notice that his countrymen were ordered to retire, not flying. Just as our line had entirely cleared the Spaniards, the smoke of battle was for one moment blown aside, by the slackening of the fire, and gave to our view the French grenadier caps, their arms and the whole aspect of their frowning masses. It was a grand, but a momentary sight; a heavy atmosphere of smoke enveloped us, and few objects could be discerned at all – none distinctly. The best soldier can make no calculation of time, if he be in the heat of an engagement; but this murderous contest of musketry lasted long. At intervals a shriek or a groan told that men were falling around me; but it was not always that the tumult of the conflict suffered me to catch individual sounds. The constant "feeling to the centre" and the gradual diminution of our front more truly bespoke the havock [sic] of death. We were the whole time progressively advancing upon and shaking the enemy. As we moved slowly, but ever a little in advance, our own killed and wounded lay behind us; we arrived among those of the Spaniards who had fallen in the first onset, then among those of the enemy. At last we were only twenty yards from their front.'

By this point Hoghton and one of the battalion commanders had been killed, the other two battalion commanders had been wounded, and in all two-thirds of the brigade had been hit. The diminished British line was now shorter than the frontage of the French column, but this column was in deep disarray. Its front was a veritable rampart of dead and wounded. Of the 3,000 casualties of the V Corps this day, it seems that about 2,000 were suffered in this murderous fire-fight with Hoghton's 1,600 men and the 250 survivors of the 31st Foot. Never was the superiority of line over column in a fire-fight, or the quality of British infantry, so clearly and drastically demonstrated as at Albuera.

The Battle of Albuera

For some time neither commander interfered in the slaughter. Soult watched numbly as his infantry was destroyed and did nothing to aid them, even though he had Werle's brigade, the combined grenadiers and his superior cavalry in hand and uncommitted.

Soult's despatch to Napoleon after the battle revealed that he had learned during the conflict, from Spanish prisoners, that Blake's army was indeed present on the field and that he was facing 30,000 men. He states that the odds 'were not fair' and that he cancelled any offensive action, limiting himself to a defensive stance. This was an unexpectedly timid reaction to the situation by an experienced Marshal of France.

Beresford only had the 4th Division, the three brigades of Hamilton's and Collin's Portuguese and 4,000 Spaniards that he could throw into the struggle. However, Beresford feared to move the 4th Division from its position where it provided a bulwark against Latour-Maubourg's great, threatening mass of cavalry and instead called Hamilton's 4,800-strong division down from behind Albuera to join Hoghton's brigade. But it took much longer than he expected for these men to march down from the hill by Albuera. The main reason for the delay was that some of the ADCs sent to find Hamilton were wounded en route. The other factor was that Hamilton, on his own initiative, had advanced towards the village in order to support Alten against Godinot's fierce attack on that place.

At last, Beresford ordered Abercrombie's brigade of Stewart's 2nd Division to form line to the left of Hoghton's brigade; it lapped around the right flank of Girard's dwindling mass and increased the slaughter. Beresford also tried to bring Carlos de España's men into the fight on the left flank of the French column, but all ranks in that formation were so fire-shy, that one of their colonels refused to advance even when Beresford grabbed him by his epaulettes and tried to drag him forward. Finally, Beresford ordered Alten's Germans to hand the defence of the village over to a brigade of Spanish troops and to come down south. This, of course, meant another considerable delay before the Germans could arrive, the village being some two miles from the site of the dramatic struggle. The crisis on the south flank passed before they arrived. They were then sent back to re-take the village; a task which they performed with the loss of about 100 casualties.

Meanwhile, Cole had been growing increasingly concerned at the rate at which the thin lines of the 2nd Division on his flank were wasting away. He sent an ADC to Beresford to request orders, but this officer was badly wounded on his way and his message was never delivered. Just then, Colonel Henry Hardinge, serving as Deputy Quartermaster-General in the Portuguese Army, rode up to Cole and urged him to advance as Hoghton's brigade could not last much longer. Cole's division had been standing idly in position for 90 minutes as the battle raged. The ADC's nudge tipped the scales. Cole was well aware of the threat posed by the French cavalry, but after a short conference with Lumley, he decided to go and join the uneven fight. As he would be vulnerable to Latour-Maubourg's cavalry, he advanced in *ordre mixte*, with a battalion in column at either end of his line. Lefebure's horse artillery battery was pushed out as a barrier to the French cavalry and all the British and Spanish cavalry under Lumley faced off against Latour-Maubourg.

Soult saw at last that he could no longer allow his V Corps to be destroyed without making some effort to extract it. He ordered Werle's brigade to advance against Myers's Fusilier Brigade of Cole's 4th Division and sent Latour-Maubourg to charge Harvey's Portuguese. Latour-Maubourg despatched four regiments of dragoons against Harvey, thinking to brush the Portuguese away as easily as he usually dealt with unsteady Spanish infantry in the open. Today was another moment of truth for the French.

Harvey's 11th and 23rd Portuguese Infantry Regiments had not seen any serious action before. This of all days was to be their baptism of fire. As the dragoons thundered down upon them they remained perfectly steady, then greeted the enemy horsemen with a series of well-aimed volleys, which utterly shattered them. The dragoons streamed away to the rear and there was no other cavalry assault for the next half hour. Myers's brigade had 2,600 men and was arranged in *ordre mixte*, with a two-deep central line and flank supports in column; the Loyal Lusitanian Legion on the left, Harvey's Portuguese on the right. Myers now advanced up a slight rise to clash with Werle's

The Battle of Albuera

brigade, which was deployed into three regimental columns of divisions. This gave the French force a frontage of about 180 men, 30 ranks deep, which was opposed by a line of 250 files who were able to use their muskets. Thus some 360 French weapons were fighting against some 500 of the allies.

Colonel Blakeney commanded the 2nd Battalion of the 7th Fusiliers (later to become the Royal Fusiliers). He wrote of the subsequent events as follows:

> 'From the quantity of smoke, I could perceive very little but what went on in my own front. The 1st Battalion of the 7th closed with the right column of the French; I moved on and closed with the second; the 23rd took the third. The men behaved most gloriously, never losing their rank, and closing to the centre as casualties occurred. The French faced us at a distance of about thirty or forty paces. During the closest part of the action I saw their officers endeavouring to deploy their columns, but all to no purpose. For as soon as a third of a company got out, they would immediately run back, to be covered by the front of their column.'

This astounding, ill-matched fire-fight continued for about 20 minutes, when suddenly, the French infantry broke and fled back up the hill, dissolving into a mass of disordered men. The Fusiliers followed, still firing, until they crested the ridge, where the fire from the 40 guns of Soult's reserve batteries put an end to their chase. The Loyal Lusitanian Legion lost 171 men out of 572; the Fusiliers lost 1,045 out of 2,015, including their commander, Myers, who was among the killed. Harvey's Portuguese had only very slight losses. Werle's brigade lost some 1,800 out of the 5,600 present.

At the same time that the Fusiliers advanced against Werle's great column, Abercrombie's brigade came up on the left of the remnants of Hoghton's brigade and moved on against the right flank of the mess that was Girard's V Corps, gave them a volley and charged. The French could take no more. They broke and ran, mingling with Werle's fugitives to cross the brook to safety. The allied artillery caused havoc in the last formed French infantry, the two battalions of combined grenadiers, who lost 372 men out of 1,000 in 20 minutes. Wisely, Beresford decided not to press on with a pursuit; he was still facing Soult's artillery, some 40 pieces, and his unshaken cavalry, which far outnumbered that of the allies. A heavy, drenching rain now began and ended this momentous day.

The Battle of Albuera had underlined two truths. Firstly, it was one of the most convincing demonstrations of the superiority of British infantry tactics over those of the French. Against all expectations, outnumbered British lines had repeatedly, and devastatingly, crushed two massive French columns and caused them crippling casualties. The combat quality of the allied soldiers involved, Zayas's Spaniards, Harvey's Portuguese and the British troops of Stewart's and Cole's divisions, was above reproach; they all fought and died like veterans. The battle had also shown just how destructive even a small body of cavalry could be, given the right opportunity.

Beresford's men and their Spanish allies spent a miserable night, but at least they could be comforted with their victory. The morale of most of Soult's men, particularly in the shattered regiments of V Corps, was rock bottom. Neither army moved that night. Next morning, both formed line of battle – just in case – but there was no further action, though the main body of Kemmis's brigade, 1,400 men, arrived during the day from Badajoz. Soult withdrew on 18 May, abandoning several hundred wounded in the woods due to lack of transport.

Soult withdrew in two groups; General Gazan, with 2,000 men from the regiments which had been most severely damaged, escorted the transport, the wounded and about 500 British prisoners back along the main road to the south. Over 200 of these prisoners managed to escape and rejoin their regiments in the next four days as their guards were too exhausted to keep proper watch. Soult and the remaining 14,000 men took a more circuitous route to the east via Solana and Fuente del Maestre, then south towards Llerena and through the Sierra Morena to Seville. Latour-Maubourg's strong cavalry rearguard kept their outnumbered allied opponents at a respectful distance. By dawn on 19 May, Badajoz was again shut in by Beresford's infantry.

Albuera was the bloodiest battle of the entire Peninsular War; it was two days before the last surviving wounded were brought in. The losses of both sides were amazingly similar and

The Battle of Albuera

may be seen in detail in the appendices. Beresford lost 5,916, Soult about 5,936. The distribution of these casualties among the various regiments present was most uneven. As always, it is very difficult to obtain reliable data on French battle casualties.

Soult, as was common practice among French Napoleonic commanders, lied in his despatch on the battle, admitting only to '2,800 killed and wounded', and '262 officers killed, wounded and missing'. Sir Charles Oman later established the truer figures. The conduct of neither commander was perfect, it seldom is. Sir William Napier was scathing in his criticisms of Beresford's conduct, but much of what he wrote was inaccurate and he was a bitter personal foe of the marshal and this clouded his judgement. A large portion of the blame for the high British casualties must lie with General Joachim Blake's disregard of Beresford's instructions to reinforce the threatened wing promptly and sufficiently. Many contemporary observers, and later historians, assess Blake as being one of the best of the Spanish generals, but 'unfortunate'. There would seem to be the possibility that Blake was one of the chief architects of his own misfortune.

Born in Malaga of Irish descent in 1759, he was a colonel in 1808 when he was appointed Captain-General of the armed forces in Galicia. In actions against the French he commanded under Cuesta at the defeat at Medina del Río Seco, but could not be said to have been primarily responsible for that disaster. He commanded, and was also defeated, at Zornoza and Espinosa and was subsequently replaced by La Romana. In 1809 he was re-employed as commander of the army in Aragon and later in Catalonia. On 23 May 1809 he defeated Suchet at Alcañiz and caused him to abandon most of Aragon for a time. Blake's handling of his army at his defeat at Maria del Huerve on 14/15 June 1809 was extremely poor; he was then defeated at Belchite two days later. When he failed to relieve Gerona, he resigned in December 1809. The available selection of Spanish generals was of such quality at that time that Blake was again offered command in time to turn up at Albuera and wreak his old magic again. Following this near-catastrophe, he was given the post of Captain-General of the armies of the Centre, Aragon and Valencia. Eventually he was cornered in Valencia, which surrendered on 9 January 1812 after a sloppy defence under his command. Blake was sent to captivity in Vincennes, near Paris, until 1814. In later life he blossomed as a liberal politician; he died in 1827.

The real villain among the allies on this day was General Sir William Stewart, commander of the 2nd Division. He allowed himself to be so swept along by the excitement of the action, that he threw his men into the fight willy-nilly and ignored Colonel Colborne's request to form a block against the imminent danger of an enemy cavalry attack on his right flank. Colborne subsequently had a very distinguished career and was instrumental in the defeat of the final French assault at Waterloo. The real victor of Albuera was Cole, of the 4th Division who, without any orders from Beresford, did what had to be done to avert an allied debacle and to allow a near-defeat to be re-forged into a great-if-bloody victory. Beresford was shocked by the losses suffered this day by the troops under his command, and had to be reassured by Wellington to carry on in his position of command.

Wellington rode over the battlefield on 21 May 1811. He wrote:

> 'We had a very good position, and I think should have gained a complete victory without any material loss, if the Spaniards could have manoeuvred; but unfortunately they cannot.'

Also:

> 'The Spanish troops behaved admirably, I understand. They stood like stocks while both parties were firing into them, but they were quite immovable, and this was the great cause of all our losses. After they had lost their position, the natural thing to do would have been to attack it with the nearest Spanish troops, but these could not be moved. The British troops (2nd Division) were next, and they were brought up (and must in such cases always be brought up), and they suffered accordingly.'

On the French side there were two main culprits for their defeat: Soult and Girard. Until Soult received the news that Blake's army was already on the field, his plans had worked

The Battle of Albuera   93

perfectly and an entire enemy brigade had been destroyed for little loss. Although his V Corps was wasting its time and being shot to pieces, he could have saved the situation by ordering prompt support of it by a determined cavalry assault. Instead, he seems to have been so shocked by the 'unfair odds' which he unexpectedly now faced, that his mind was paralysed for some time. This delay was fatal for his army. Soult built up a solid reputation as being repeatedly outwitted by Wellington in the Peninsula, and an equally robust one as being an expert looter for his own personal benefit. In 1815 he embraced the Restoration and became Louis XVIII's Minister for War and was active in persecuting Bonapartists. In the Hundred Days he returned to Napoleon's fold to act controversially as his chief of staff. Following the Second Restoration he became a Peer of France. In 1847 he was appointed Marshal General of France. He died in 1851.

The other main culprit was Girard, who chose to form V Corps into a massive column and failed to deploy it properly before it was ground to pieces in the British line's mincing machine. On 28 October 1811 his division was ambushed at Arroyo dos Molinos and destroyed by General Hill's 2nd Division. Girard was recalled to France in disgrace. In 1812 he commanded the 28th Division, IX Corps, in Russia and was wounded at the Beresina. In 1813 he was badly wounded at Lützen, then defended Magdeburg successfully until Napoleon's abdication. In the Hundred Days he rallied to Napoleon, commanded the 7th Division, II Corps, and was mortally wounded at Ligny, dying on 27 June 1815.

Amazingly enough, neither Soult, nor any other French officer present, seems to have had any tactical pennies drop as a result of this dramatic day. As late as nightfall at the end of the battle of Waterloo, no French commander could think fit to deploy his massive columns of infantry into line before they were shot to shreds, 'in the same old way', by the British.

But the mightiest event had been the cavalry assault on Colborne's brigade. The results were rapid and amazingly destructive, cavalry at its best. Within the space of a few minutes, Colborne's four battalions were cut down, the 3rd Foot (The Buffs) losing 643 men out of 755. Of these no fewer than 216 were killed against 248 being wounded. This extremely unusual high ratio of killed to wounded in Colborne's brigade is a witness to the savagery of the Poles, who not only often refused to take prisoners, but stabbed the wounded as they lay on the ground. Their conduct was unacceptable by any standards. Not content to behave so savagely in the heat of actual combat, the Poles seemed to be completely out of control for the entire day. Major Brooke, commanding officer of the 2nd Battalion, 48th Foot, had been captured and was being led to the rear by two French soldiers, when a Polish lancer rode up and cut him down. Brooke's account of this event is of interest.

> 'Part of the French cavalry were Polish lancers: from the conduct of this regiment on the field of action I believe many of them to have been intoxicated, as they rode over the wounded, barbarously darting their lances into them.
>
> Several unfortunate prisoners were killed in this manner, while being led from the field to the rear of the enemy lines.
>
> I was an instance of their inhumanity: after having been most severely wounded in the head, and plundered of everything that I had about me, I was being led as a prisoner between two French infantry soldiers, when one of the lancers rode up, and deliberately cut me down.
>
> Then, taking the skirts of my regimental coat, he endeavoured to pull it over my head.
>
> Not satisfied with this brutality, the wretch tried by every means to make his horse trample on me, by dragging me along the ground and wheeling his horse over my body.
>
> But the beast, more merciful than the rider, absolutely refused to comply with his master's wishes, and carefully avoided putting his foot on me!
>
> From this miserable situation I was rescued by two French infantry soldiers, who with a dragoon guarded me to the rear.
>
> This last man had the kindness to carry me on his horse over the river Albuera, which, from my exhausted state, I could not have forded on foot.'

The British admitted the loss of five of the six colours of Colborne's brigade (one of the 3rd Foot, two each from the 48th and 66th); all French and Polish accounts claim six, but five

The Battle of Albuera

seems to be the real figure. Five guns of Cleeve's battery were overrun in the charge, although only one howitzer was actually taken off. And all this was achieved by two cavalry regiments. Of the British colours lost, one (that of the 48th Foot) was taken by Sergeant Dion d'Aumont of the 10th Hussars. The other four were taken by the Lancers of the Vistula Legion. Apparently they survived the general destruction of trophies ordered by the French in 1814, and turned up again in the Artillery Museum in 1827. In the revolution of 1830, the regimental colour of the 66th Foot was looted by the mob. On 26 February 1861 General Dufforc d'Antiste donated his collection of trophies to the Invalides. This included the colour of the 66th, which he had bought in the intervening years. The other four were hung in the church of Les Invalides, where, in a fire which started during the funeral of Marshal Sebastiani on 11 August 1851, they were further damaged. The king's colour of the 66th was destroyed completely; the king's colour of the 48th and scraps of the regimental colours of the 3rd and 48th survived. They are held in Les Invalides.

French cavalry losses at Albuera were severe. The 2nd Hussars lost about 24 per cent of their men. According to the French returns used by Oman the Lancers of the Vistula lost 130 men (other sources suggest a figure of about 150), some 22–26 per cent of their strength of 591 at the start of the day. Lieutenants Rogoyski and Wojciechowski were awarded the cross of the Legion of Honour. Colonel Konopka was promoted to general on 6 August 1811. He was later colonel-major of the 1st Polish Chevaux-légers-Lanciers of the Imperial Guard, and in January 1812 major-colonel of the 3rd Lithuanian Chevaux-légers-Lanciers, also of the Guard. This regiment was largely destroyed by the Russians at Slonim on 19 October 1812 and Konopka was captured. He settled in Warsaw after the war.

Losses notwithstanding, almost directly after the bloody battle of Albuera Napoleon ordered the conversion of six regiments of dragoons and the cavalry of the Hanoverian Legion to lancers.

Sources

Hall, John A., *History of the Peninsular War*, Vol VIII, London, 1998

Kirkor, S., *Legia Nadwislanska 1808–1814*, Warsaw, 1982

Martinien, A., *Tableaux pars Corps et pars Batailles des Officiers Tués et Blessés Pendant les Guerres de l'Empire 1805–1815*, Paris, 1890

Oman, Sir Charles, *History of the Peninsular War*, Vols I–VII, London, 1996

Oman, Sir Charles, *Studies in the Napoleonic Wars*, London, 1989

The scene at the point of Kellermann's decisive charge at the climax of the Battle of Marengo, 14 June 1800. From the painting by Bellange, 1832.

General Bessières leads the cavalry of the Consular Guard (Grenadiers- and Chasseurs-à-Cheval) into action in the final stages of the Battle of Marengo.

General der Kavallerie Michael Friedrich Benedikt Freiherr von Melas. At the time of Marengo Melas was 71 years old, an incredible age for a soldier on active duty.

General-Major Anton von Zach who was captured during Kellermann's famous charge at Marengo.

General Louis Desaix. His death in the hour of victory at Marengo helped Napoleon claim the credit for the triumph even though Desaix had been its true architect.

Charge of the Grenadiers-à-Cheval of Napoleon's Imperial Guard at the Battle of Austerlitz.

Charge of the Mamelukes at Austerlitz.

Feldmarschall-Leutnant Michael Freiherr von Kienmayer, Commander of the Austrian Advanced Guard of the 1st Column at Austerlitz.

Feldmarschall-Leutnant Prince Johann von Liechtenstein, commander of the Austrian forces at Austerlitz. His large decoration is the Commander's Order of Maria-Theresa.

General Rapp presents captives and colours of the Russian Imperial Guard to Napoleon at the end of the Battle of Austerlitz.

Napoleon and his entourage view the scenes of carnage at Eylau. Napoleon's official bulletin greatly understated the casualties of this bloody battle.

The Battle of Eylau, 8 February 1807, from an aquatint by J.L. Rugendas.

Marshal Davout as Colonel-General of the Chasseurs-à-Pied of the Guard.

Marshal Soult as Colonel-General of the Grenadiers-à-Pied of the Guard.

Above: The highly successful Cossack commander, Count Matvei Ivanovich Platov.

This shows Uvarov at the most extreme point his raiders reached on the northern flank of Borodino on 7 September 1812.

Right: Russian Cuirassiers clash with Saxon Garde du Corps in the struggle for the Raevski redoubt; Borodino, 7 September 1812.

Left: General of Cavalry Baron Levin August Theophil von Bennigsen, commander of the Russian Army at the Battle of Preussisch-Eylau. It was Bennigsen who recommended the adoption of the deep Russian infantry formations at Borodino which provided Napoleon's artillery with such wonderful targets.

Above: Monument to Uvarov's raid, showing the open terrain in the area, ideal for cavalry movement.

Right: Crossing the Beresina, a howitzer shell causes consternation in the French camp, from a lithograph by Emminger after Faber du Faur.

Below: Borodino Church. This chapel still stands today as it was in 1812.

Right: Don Cossacks; drawn from life by Georg Adam in Nuremberg in 1814. An interesting roadside study of the much feared light horsemen and their hardy ponies. Troops of the corps caused the Grande Armée much trouble in Russia in late 1812.

Blücher's cavalry in action during the ambush at Haynau, 26 May 1813. From Müller-Boehm's *Die Deutschen Befreiungskriege*.

Prince Eugène as Colonel-General of the Chasseurs-à-Cheval of the Guard.

Feldmarschall Gebhard Lebrecht Fürst Blücher von Wahlstadt. Born in 1742 and originally in Swedish military service, Blücher was an old warhorse of 71 in 1813. No one could ever accuse Blücher of military genius, but he loved active service and could inspire his men to about the same degree that Napoleon could. He was the architect of the successful ambush of Count Maison's 16th Division at Haynau on 26 May 1813.

French hussars in action at Leipzig. From *Die Deutschen Befreiungskriege*.

A mêlée around a French eagle in the last, desperate phase of the Battle of Leipzig on 19 October 1813. Austrian (Hungarian) grenadiers attack from the left, while cavalry charge in from the right. The design of the French colour would appear to be fanciful.

Franz I, Emperor of Austria. This portrait shows the Hapsburg leader after his renunciation of the throne of the Holy Roman Empire in 1806.

King Frederick William III of Prussia. His indecision had resulted in Prussia being defeated and dismembered by Napoleon's Grande Armée in 1806.

Field Marshal Fürst Karl Philipp zu Schwarzenberg. Nominally Allied Commander-in-Chief in 1813–14, Schwarzenberg had to operate under direct scrutiny from the Prussian, Austrian and Russian monarchs.

General Freiherr von Langenau, Quartermaster-General to Prince Schwarzenberg in the Leipzig campaign in 1813. Langenau had served as Chief of Staff of the Saxon contingent in Russia in 1812.

Prussian cavalry in action during the Battle of Möckern on 16 October 1813.

Left to right: Cossacks, Bashkirs and Kalmucks, drawn from life by Georg Adam in Nuremberg, 1814. These hardy horsemen formed the main component of many of the Allied raiding parties of the 1813 campaign. Contemporary accounts confirm that the Bashkirs and Kalmucks were largely armed with bows and arrows or lances.

Russian hussars of the 1812–14 era in action. Note the Kiver; apparently the Mariupol Regiment is shown here.

French Young Guard in action against Russian hussars in the 1814 campaign. The youth of the 'Marie-Louises' is obvious. Despite this, they often fought as well as veterans.

General Colbert leads the French Guard Lancers against a British square at Waterloo.

2nd Life Guards at Waterloo, June 1815. Watercolour by Richard Simkin, *circa* 1900.
[Anne S.K. Brown Military Collection]

6th Inniskilling Dragoons at Waterloo. Watercolour by Richard Simkin, *circa* 1900.
[Anne S.K. Brown Military Collection]

CHAPTER 6

The Clash at García Hernandez
23 JULY 1812

In the previous chapter, the Battle of Albuera showed us what cavalry could do to infantry caught in the open in line. This action features a series of cavalry attacks against infantry in square.

Throughout the Napoleonic era, and beyond, the classical defensive form adopted by infantry caught in the open by the threat of a cavalry attack was the square. This could be hollow or solid (as in the case of the Austrian 'Mass') and formed by anything from a battalion up to a division. It was often oblong rather than exactly square and could be formed in a variety of ways. The advantages of the square were that it presented no vulnerable flanks to the predatory cavalry and at the same time allowed the optimum firepower to be brought to bear on the enemy. It was a formation which could be used in a mobile as well as a static role. An example of the former was shown on 6 September 1813 in the French defeat at Dennewitz, where the Württembergers of the 2nd and 7th Infantry Regiments withdrew in squares for almost a mile over open country under heavy fire. The main disadvantage of square was that it presented the enemy artillery with an excellent target. When formed by steady infantry, a square was almost invulnerable to cavalry attack, unless infantry and artillery were also brought into play against it. Examples of squares being broken were rare in the extreme.

If the breaking of any square was rare, the destruction of two on the same day was a singular event. That this feat should be achieved by two cavalry regiments undergoing their baptisms of fire makes the achievement even more remarkable. Yet this was done by the 1st and 2nd Dragoons of the King's German Legion (KGL) in Spain on the day after Wellington's skilful victory over Marmont at Salamanca in 1812.

The facts of this great feat are as follows. Prior to the battle of Salamanca on 22 July, Wellington had finally secured the key Spanish frontier fortresses of Badajoz and Ciudad Rodrigo and destroyed the bridge at Almarez on the Tagus, in preparation for his invasion of northern Spain to expel the French. By 27 June he had taken the forts and town of Salamanca, after a ten-day siege, which his immediate opponent, Marshal Marmont, commander of the Army of Portugal, had not dared to attempt to break. The strategic situation of King Joseph's French armies in their uncomfortable occupation of his hostile realm had become so weak at this point that no effective, aggressive response could be mounted against Wellington's advance. Too many French resources had been drained off to die in Napoleon's great gamble in Russia to prop up his brother's troubled regime in the Spanish sideshow.

Marmont's army was now streaming away from their decisive defeat at Wellington's hands in the battle of the previous day. The French know this defeat as the Battle of the Arapiles, the actual site, which was about six miles south of Salamanca itself. The action had been fought after weeks of operations in which both commanders had minced carefully around one another in northern central Spain, some 110 miles north-west of the French-held capital of Madrid. Finally, Wellington detected an error by his enemy and struck. The victory had cost the French some 14,000 casualties from their 43,000-strong army, along with 20 guns, two eagles and six colours. Marmont himself had been wounded as had Generals Bonnet and Clauzel; Generals Ferey and Thomières had been killed. Some 10,000 more of the surviving French soldiers were straggling, in disordered leaderless groups, many without weapons, away to the south-east, towards Peñaranda and Madrid.

By pure chance, Wellington's pursuit of the beaten French

The Clash at García Hernandez   115

was misdirected and lacked speed. The panic-stricken fugitives were thus able to cross the River Tormes unhindered and to make their way off to safety that night.

One French officer, Lemonnier-Delafosse, of the 31st Light in Ferey's division wrote of what he saw on the evening of 22 July:

> 'A shapeless mass of soldiery was rolling down the road like a torrent – infantry, cavalry, artillery, wagons, carts, baggage mules, the reserve park of the artillery drawn by oxen – were all mixed up. The men, shouting, swearing, running, were out of all order, each one looking after himself alone – a complete stampede. The panic was inexplicable to one who, coming from the extreme rear, knew that there was no pursuit by the enemy to justify the terror shown. I had to stand off far from the road, for if I had got near it, I should have been swept off by the torrent in spite of myself.'

Had Wellington chased them with any vigour, the wreck of Marmont's army would have been trapped against the River Tormes and forced to surrender. Instead the fugitives scattered away to the south-east, using three roads; the bulk of the army using that leading through Peñaranda. Foy's division formed the rearguard at Alba de Tormes early on 23 July. When Anson's light dragoon brigade appeared on the far side of the river, Foy's artillery fired a few obligatory rounds, then the whole division decamped to join its comrades. Wellington was with Anson and ordered patrols to follow down each road, but concentrated his main effort on that leading to Peñaranda. About seven miles on from Alba de Tormes, they came upon Foy's division again, behind the Caballero brook, north of the small village of García Hernandez. As soon as he saw the British appear, Foy resumed his withdrawal; French officers were sent into García Hernandez to collect up the crowds of stragglers there who were getting water.

Curto's light cavalry division was drawn up on a low ridge, north of the village, to hold up the chase. This ridge concealed Foy's infantry from Wellington's view. As he withdrew, Foy's order of march was as follows. 39th Line, 69th Line, 1st Battalion, 76th Line, 6th Light, 2nd Battalion 76th Line, and finally Curto's cavalry with a horse battery. At this point,

Anson's brigade was alone; the infantry of the 1st and Light Divisions were some miles to the rear. Anson had two squadrons each of the 11th and 16th Light Dragoons under his immediate command (though an article in *Blackwood's Magazine* of March 1913 states that the 12th Light Dragoons were also present), and had sent von Bock's 1st and 2nd Dragoons of the King's German Legion off to the north, to deliver a left hook into the enemy's flank. As the British light dragoons advanced, Curto's regiments, still unsteady after the defeat of the previous day, fled away to the rear, not awaiting the shock. They thus escaped von Bock's planned left hook.

Major-General Eberhard von Bock, an extremely short-sighted officer, was leading his brigade down the road from Huerta, over the River Almar, when an officer of Wellington's staff galloped up to him. It was Lieutenant-Colonel John May of the Royal Artillery, with the order for von Bock to charge the enemy cavalry. Von Bock peered myopically into the distance and saw just a dusty blur. He asked May kindly to point out his target for him. May, ever ready for a bit of action, did better than that and offered to ride with von Bock to ensure that he charged home at the enemy as planned. In the following action, May was wounded several times. For this act of derring-do, he was in Wellington's bad books for some months; the duke frowned upon officers exposing themselves needlessly to risks. In later life, May would retell the tale and add: 'and that's what I got for playing the dragoon and leading the Germans against the enemy!'

While von Bock and May charged after Curto's division with the 1st KGL Dragoons, the 2nd KGL Dragoons advanced on their left, into an area they thought to be empty of the enemy. The *Blackwood's Magazine* article states that the Dragoons chased off the 15th Chasseurs-à-Cheval and the 1st Gendarmerie at this point, after a sharp action.

The German general could not see exactly what was going on, and Anson and Wellington were out of sight, so the subsequent dramatic victory was entirely the work of the two leading squadron commanders of the 1st and 3rd Squadrons, 2nd Dragoons. This happened when they took fire in the left flank as they chased after the French cavalry. This came from the

The Clash at García Hernandez

rearmost square of the 76th Line, which the dragoons had not noticed until then, and which was invisible to all the allied commanders. The 76th Line and the 6th Light, about 2,400 men in all, were the tail end of Foy's infantry and had been unable to make good their escape. Three of these battalions were in square and one battalion of the 6th was hurrying off in column.

The rear square had delivered its first volley at a range of about 90 yards. This brought down several men and horses, and wounded Captain Gustavus von der Decken (commanding 3rd squadron, 2nd KGL Dragoons), though he managed to stay in the saddle for the moment and continued to lead the charge. At a range of 20 yards, the square fired again. This destructive volley caused von der Decken to fall, but did not break the speed or the resolve of the Germans. They rushed on at the square. A wounded horse, that of Trooper Post, reared up and crashed into the French ranks, kicking and rolling, causing a breach in their wall. Several dragoons seized the chance and jumped into the interior of the square through the gap thus created. Now the already shaky infantry had had enough. The square dissolved in disorder. Several men were cut down but most just threw down their weapons and surrendered. Only about 50 escaped. Later that day, eye-witnesses recorded seeing French muskets laid out in a neat square on the site. Sixteen officers of the 76th were captured, only two of whom were wounded. One officer was killed and five of the other officers shown by Martinien as having been wounded in this action did not reach captivity in England and probably died of their wounds. Gustavus von der Decken died in Salamanca on 16 September 1813.

The action at García Hernandez went on at a racing pace. To the right of the broken square, the two battalions of the 6th Light, now in column, were scrambling up the side of the hill, trying to escape their doom. The dragoons could hear the officers shouting out: 'Allongez le pas, gagnons la hauteur!' ('Lengthen your stride, let's get to the top.')

Captain von Reitzenstein's 2nd Squadron, 1st KGL Dragoons, was closest to the French and he ordered his men into a charge at once. Just before they hit the column, the two rearmost companies (the Carabiniers and the 1st Company, 1st Battalion, under Captain Philippe) faced about and greeted them with an effective volley at 50 yards. Then the cavalry were among them and the slaughter began. The dragoons rode along the sides of the straggling column; the execution was great.

At this point, the 2nd KGL Dragoons also arrived on the scene. They, and some of the 1st KGL Dragoons went after the mass of French infantry, the other battalion of the 76th, which was desperately trying to form square as the fugitives of the 6th flooded into it. The unfinished square collapsed under the charge of the dragoons. Those that could, escaped and fled after their luckier comrades. At the same time, Captain von Marschalk, now commanding 3rd Squadron, 2nd KGL Dragoons, supported by Lieutenant von Fuemetty's troop of the 2nd Squadron, came up and drove off some of the enemy cavalry who had rallied and were now attempting to mount a counter-attack. These two squadrons then went on to scatter a further square that the fugitives were trying to form.

Now utterly disorganised, but drunk with victory, the Germans tried for the hat trick and rode against the next square, that of a battalion of the 69th Line. This formation had taken up the survivors of the 6th Light and General Baron Chemineau, commander of the 1st Brigade of Foy's division. Three other French battalion squares were in the same location and by now Foy had managed to bring some artillery into action. The square of the 69th held firm and delivered an effective volley, which killed Captain Friedrich von Uslar (of the 2nd Dragoons), who was leading the charge, and brought down many of his men. Von Fuemetty was brought down as well and was lightly wounded.

The mad career of the dragoons was over; their order was gone, their horses blown, their numbers reduced, the enemy too numerous. They pulled away to the side and cantered back to collect themselves and to count the costs of this remarkable day.

Senior Surgeon Detmer of the 2nd KGL Dragoons was present with his regiment and described the events as follows:

> 'At 1 o'clock in the morning of 23 July 1812, the brigade was ordered to break camp and to move off in pursuit of the enemy together with Anson's brigade, the 1st and the Light Divisions. Due to having to water the horses and to wait for

The Clash at García Hernandez

the infantry, it was 8 o'clock before we reached the right bank of the Tormes. At about 9 o'clock, Anson's brigade, which was ahead of us, saw the enemy in the plain before García Hernandez and charged at them. When we could see the enemy, which consisted of cavalry and infantry, the regiment formed up.

Anson's brigade passed right, behind García Hernandez and charged the French cavalry, which fled, and we were ordered to take the enemy in flank. We went off at a smart gallop. The 1st Squadron went ahead and charged the cavalry, which fell back; they took five horses and an ammunition wagon pulled by four mules.

The 3rd Squadron, led by Captain Gustavus von der Decken, charged off without being ordered to, against a solid infantry square on a hill to the left and broke it, and captured them all, despite being greeted with terrible musketry fire.

Captain von Reitzenstein followed them with the 2nd Squadron and charged a second square, higher up the slope, and was also lucky enough to break in and capture it.

The prisoners – apparently between 1,400 and 1,500 – were then brought in.

In this action, which was so honourable for the regiment, particularly for the 3rd and 2nd Squadrons, Lt Voss, Lt Huegel, one NCO and 27 men were killed; Captain von der Decken, Cornet Tappe, three sergeants and 31 men were wounded. After the wounded had been brought into García Hernandez, the regiment – which was very weak – resumed the advance, but received the order to set up bivouac next to the village.'

Another eyewitness was Captain Carl von Hodenberg of the 1st KGL Dragoons. On 25 July 1812 he wrote the following letter home:

'Your friends survived the battle of Salamanca on the 22nd unhurt. On the 23rd our brigade covered itself with glory. I am proud of the reports that you will read in the newspapers. We broke three formed French infantry squares (a feat that has not been achieved to date by the cavalry). General Bock led us in great style. Decken's squadron was the first, Reitzenstein's the second; our great deeds are due to the

unforgettable heroism of these two – but apart from these, everyone showed great bravery, otherwise we could never, never have achieved such success. We stormed forward like madmen, our swords held high to exact that bloody revenge that we had owed the enemy for so long. In less than half an hour we took 1,500 prisoners, and 200 more by evening. Lord Wellington was there; he declared that he had never before witnessed such a charge. The French had taken post on a ridge and we shot up out of a sunken road – our losses are very heavy (my heart bleeds that I have to write this); Captain Uslar of the 2nd (not ours), Lieutenant Voss, Heugel and Cornet Tappe [Tappe in fact survived the battle] were killed as they broke into a square. Captain Decken and two other officers are badly wounded; 150 men and horses are dead or so badly wounded that we could lose them at any time. Reitzenstein is healthy as is the general and I, and we survivors are overjoyed at our victory. Half of our brigade are casualties, General Bock now commands all the cavalry as Sir Stapleton Cotton was wounded in the battle of Salamanca. I have nothing here with which to seal this letter,
Your Servant
C. Hodenberg.
Excuse the haste, but we have nothing but our lives.'

Indeed, the dragoons had suffered 127 casualties out of the 760 all ranks present in the two regiments. The 1st Regiment lost 2 officers and 28 men killed, 2 officer and 37 men wounded. The 2nd Regiment lost 1 officer and 21 men killed, 1 officer and 29 men wounded. Six men were missing. But the destruction that they had inflicted on the enemy was out of all proportion to their own losses. An entire battalion of the 76th had been captured; of the 27 officers present, 1 was killed, 5 wounded, 16 captured. The two battalions of the 6th Light had been almost destroyed. Its commander, Colonel Molard, and 6 other officers had been taken prisoner, 8 others had been wounded. Some 500 men had also been captured. Total French losses for the day were about 1,100 men, altogether a devastating demonstration of the power of cavalry under the right circumstances, but unfortunately Wellington's failure to press the pursuit harder allowed Foy to escape with what was left of his division to fight another day.

The Clash at Garcia Hernandez

Foy, in his *Guerre de la Peninsule*, freely admitted that this charge was one of the most dashing and effective of the whole war in Spain and Portugal. And it was all the result of the initiative and lightning decisions of the commanders of the leading squadrons. Both regiments were granted the battle honour 'Garcia Hernandez', which their descendant regiments in the Hanoverian Army continued to wear on their guidons. Neither of the British cavalry regiments that were in the same area that day was awarded the battle honour.

Sources

Blackwood's Magazine, March, 1913

Dehnel, H., *Erinnerungen deutscher Officiere in britischen Diensten aus den Kriegsjahren 1805–1816*, Hannover, 1864

Hall, John A., *A History of the Peninsular War*, Vol VIII, London, 1998

Hodenberg, Karl von, *Briefe des Rittmeisters Karl von Hodenberg*, published in Schwertfeger, Bernhard, *Geschichte der Königlich Deutschen Legion 1803–1816*, see below

Martinien, A., *Tableaux pars Corps et pars Batailles des Officiers Tués et Blessés Pendant les Guerres de l'Empire 1805–1815*, Paris, 1890

Oman, Sir Charles, *A History of the Peninsular War*, Vol V, London, 1996

Schwertfeger, Bernhard, *Geschichte der Königlich Deutschen Legion 1803–1816*, 2 Vols, Hannover and Leipzig, 1907

CHAPTER 7

The Battle of Borodino
7 SEPTEMBER 1812

The Russian Raid on the Northern Flank
Whilst most of the cavalry actions in this book involve participants in the midst of large, set-piece battles, this example illustrates the potential of *ad hoc* operations, undertaken on the initiative of an imaginative commander. It was a side show of a massive battle; one of the bloodiest of the Napoleonic era. Had it been more appropriately supported by the Russian high command, it might well have caused Borodino to be recorded in history as a crushing Russian victory. It might well also have terminated Napoleon's career three years earlier than actually happened. As it was, this raid on Napoleon's unprotected northern flank saved the Grand Battery from final capture for about two hours.

At about 7:30 on the morning of the battle, Cossack patrols sent out by Ataman Platov (commander of the Don Cossack *pulks* operating on the extreme northern Russian flank) discovered a ford across the lower Kolocha in the area of the hamlet of Selo Novi. Seeing that the terrain west and north of the stream was empty of allied forces, Platov sent one of his aides, Colonel Prince Ernst von Hessen-Philippsthal (who was serving with the Russian Army in a voluntary capacity) to Kutuzov's headquarters to ask permission to mount a raid into this vacuum to see what damage might be done to the enemy. On his way to Kutuzov at Gorki, the prince met another German, the tireless and imaginative Colonel von Toll of Kutuzov's staff. After discussing the projected raid, Toll undertook to recommend the idea to the Commander-in-Chief, which he at once did. Toll suggested that not just Platov's Cossacks, but also General Uvarov's I Cavalry Corps should be committed to the action. Kutuzov agreed, saying simply: 'Eh bien! Prenez le!' ('Very well! Take it!'). French was commonly spoken in most armies of the time, and there were many foreigners serving with the Russian staff in 1812.

Fortune had favoured the spontaneous plan because it was at about this time (10 o'clock) that the hotly-disputed Raevski Battery had just been re-taken from the Viceroy Eugène's IV Corps and the Russian high command were already anxious about their ability to hold out in the centre. Such a diversion might ease the pressure on their centre, and it would make good use of the available forces on their right wing.

There has been considerable criticism of this raid for its failure to achieve more than it actually did on this day. This criticism is utterly misplaced and is the result of repeated misinterpretations of the aim of the raid and of a lack of realisation of the effects it had upon Napoleon's plan for the battle. Borodino was to have been the 'Austerlitz of 1812', the crushing victory that he had been pursuing since he crossed the Russian border, the victory to gain which he had mounted a frantic, and abortive, chase and already ruined his cavalry, reduced his army to a fraction of its former strength and been drawn deep into the gloomy wastes of this vast, desolate and hostile country. The day was critical to the Emperor's final success. If he could not destroy his enemies or failed to break their will to continue the war, he would have failed in his entire gigantic gamble.

Napoleon was acutely aware of the critical importance of time in warfare, the necessity never to waste a second if his schemes were to succeed. We know from various remarks the emphasis he placed upon the subject: 'Ask of me anything but time!' 'Ground we may recover, time never!' 'I may lose ground but I will never lose a minute!'

To achieve his crushing victory at Borodino, Napoleon's forces had to destroy the Russian Army during the hours of daylight on 7 September. By the time darkness fell, his enemies had to be streaming away in total defeat, with his victorious cavalry hacking the defenceless fugitives to pieces and a throng of beaten enemy generals listening to his victorious cant as his hussars threw yet more captured colours and standards onto the heaps at his feet and his ADCs tallied up the numbers of prisoners and guns taken. For this reason, the assault had to begin at an early hour. If the Russians could hold out as an intact fighting force until dark, Napoleon would have failed. Anything which prevented his army from breaking their line this day would work immensely in their favour. It should also be remembered that this raid was a completely *ad hoc* idea, not part of the prepared plan and only mounted to take advantage of a lucky chance. There were no written orders, no targets were set, no agenda laid down and, most importantly, the force sent contained only one battery of artillery and no infantry at all. Its ability to be a serious threat was thus almost nil. At the best, it was a glorified Cossack raid, snooping around on the enemy's apparently-open left flank to see if Lady Luck would throw any juicy morsels in their path.

This raid was meant by Kutuzov, and accepted by Uvarov, as merely a feint. But let us examine just what the raid did achieve. All sources that understand the background to the raid agree that it caused such a panic on the French left wing that Prince Eugène abandoned his planned third assault on the Raevski Battery at about 12 o'clock, pulled his forces out of the front line before the works and withdrew them to the west to confront this new threat. By the time that the Russian raiding force withdrew eastwards back into their own lines and Eugène could redeploy his forces for the assault on the battery, two hours had passed. His withdrawal of the IV Corps and other forces from the front line caused a gap which had to be hurriedly filled by other allied units, thus distracting them from their primary tasks and further easing the pressure on the battered Russian centre. Napoleon had made his plan for the day but here we see these bungling, incompetent Russians stealing not only a few minutes from The Master; they stole two complete hours!

These were two hours in which the Russian high command could review existing plans and revise them if needed; two hours in which shattered Russian regiments could be reorganised or replaced in the line; two hours in which exhausted Russian infantry and artillery ammunition could be replenished; two hours in which Russian reserves could be redeployed to the most threatened section of the front; two hours (and more) in which Montbrun's II Cavalry Corps had to fill a gap in the French line, in close range of the Russian artillery in and around the Battery, and during which it suffered immense damage. And it was two hours' time lost which The Master was never to recover.

The question must also be asked: should Prince Eugène be blamed for snatching from his adoptive father's grasp, the crushing victory which he so desperately needed? For there is no doubt that the Russian Second Army had been knocked to pieces in the prolonged and bitter fighting at the fleches and at Semenovskaya and its morale had suffered a severe blow when Bagration had been wounded. Many of the Second Army's units were close to dissolution by 12 o'clock and it was already the reserves of their V Corps (the Guard) and the heavy cavalry who were holding the ring for Kutuzov in this sector. The sudden and unexpected lifting of the French assaults here, caused by the need for their forces to side-step to the north to fill the gap in the line caused by Eugène's withdrawal of his units in response to the sudden appearance of Platov and Uvarov in his flank and rear, gave the Second Army a vital respite. Further, should Platov, Uvarov and von Toll be posthumously enshrined in Russia's annals as the true – but yet unsung – heroes of Borodino and architects of Napoleon's crushing defeat in Russia in 1812?

Together Uvarov's and Platov's commands amounted to some 8,000 men and 12 guns (see appendix). The force concentrated and moved off, crossing at the ford and debouching into the flat, open area north of Borodino. Here they split into two parts. Uvarov's Corps swung to the south-west and south; Platov's Cossacks initially rode due west, crossing the upper Voina stream into a large expanse of scrub and bush and then turned south to lunge down at the Kolocha in the area of the hamlets of Valueva and Aleksinki, and thus threatening the rear of the Viceroy's IV Corps, his ammunition train, the wounded of the action so far and the Kolocha bridging site, but – more importantly – aiming directly at Napoleon's command post at the Shevardino Redoubt! The two groups of Russian cavalry were now completely separated by about a mile, thus rendering mutually supportive action difficult if not impossible; each was playing his own game.

Meanwhile, just after midday, Prince Eugène had received orders from Imperial Headquarters to undertake another assault on the Raevski Battery. Part of the Young Guard and the Reserve Cavalry were to support him. General Sorbier, commanding Napoleon's Guard Artillery, had by now noticed that the Russian centre had been reinforced by the Duke of Württemberg's 4th Division and Ostermann-Tolstoi's IV Corps. He assumed that a Russian counter-attack was imminent and took action to frustrate it. He brought up 36 guns from the Guard Artillery Park and a further 49 horse artillery pieces from Nansouty's and Latour-Maubourg's I and IV Cavalry Corps. These guns, and the IV Corps' own artillery, now opened up against Württemberg and Ostermann creating the 'hell' of crossfire so fittingly described by some of those on the receiving end of it.

Just at the moment when Eugène received his attack orders panic broke out in the rear of his corps. Uvarov and Platov had been sighted bearing down on the sparsely-protected left flank. The train and baggage personnel near Aleksinki already knew enough about what Cossacks could do to their sorts if given the chance and most fled to the rear at speed. Alarming reports spread in all directions and more reached Eugène from Delzons and Ornano on his immediate left flank. The imminent assault on the Raevski Battery was abandoned in favour of more urgent damage control to assure the survival of the IV Corps.

On the site of this action stands today a red sandstone monument to the I (Russian) Cavalry Corps in commemoration of what went on here in September 1812. Engraved on the monument is a tactical map showing the Russian version of what happened. Just south of Bessubovo was a small lake formed by damming the Voina stream as it flowed south to join the Kolocha west of Borodino. Below the dam was a watermill. Due to its deep-cut nature south of the dam, and the lake above it, the only crossing point for some distance in either direction was the track over the dam itself.

According to the memorial, at the approach of Uvarov's cavalry, the available French forces in this sector were deployed

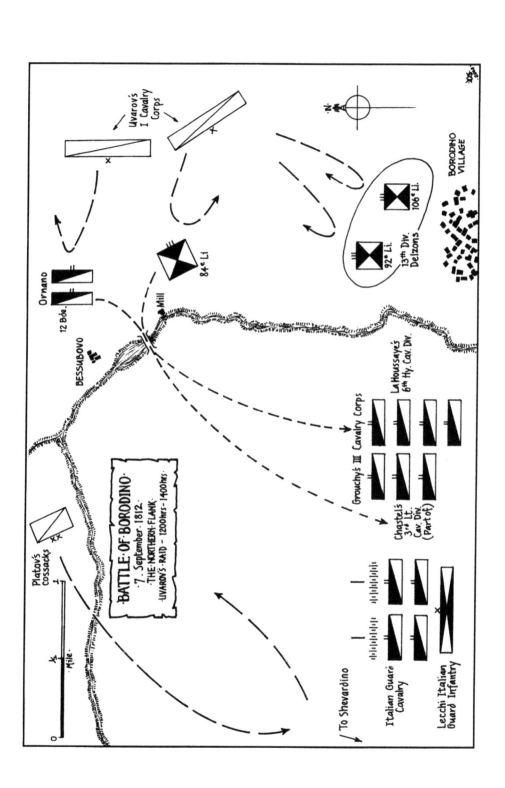

as follows: Ornano's 12th Light Cavalry Brigade (9th and 19th Chasseurs-à-Cheval) just east of Bessubovo; the 84th Line (in square) east of the mill; the 92nd and 106th Line (also in square) just north of Borodino. The location of the 8th Light is not shown but was probably in Borodino itself. West of the Voina are shown (from east to west) La Houssaye's 6th Heavy Cavalry Division (7th, 23rd, 28th, 30th Dragoons); Chastel's 3rd Light Cavalry Division (only the 6th, 8th and 25th Chasseurs-à-Cheval are shown); the Italian Guard with artillery arrayed in front (apparently four cavalry regiments – Queen's Dragoons, Guard Dragoons and possibly the 2nd and 3rd Italian Chasseurs-à-Cheval, although two of these 'cavalry' regiments might represent the two regiments of the Italian Guard Infantry which also seem to have been present here. The 3rd, 4th, 5th and 6th Bavarian Chevauxlégers Regiments were also present but are not shown on the monument.

The Prussian Colonel von Clausewitz was with Uvarov in his capacity as chief of staff, I Cavalry Corps. He suggested softening up the three French infantry squares with artillery fire but Uvarov rejected this as a waste of time. The Life Guard Hussars charged the steady 84th three times and were beaten off with loss each time. At last, Uvarov ordered the Russian horse artillery to unlimber. By this time, the 84th had withdrawn over the mill dam, but had to leave behind their two regimental artillery pieces. Further Russian charges against the squares of the 92nd and 106th were also fruitless. The total lack of an infantry component in the Russian force and their very weak artillery support ruled out any serious action against Eugène's IV Corps which was now concentrated and ready to oppose any advance across the mill dam. The weak Russian bolt was shot.

Meanwhile, Platov had crossed the Voina at a ford above Bessubovo and was well to the west of Eugène's IV Corps, spreading panic in the rear echelons. When Napoleon was informed of this unexpected Russian thrust, he was ripped out of the torpor in which he had spent much of the day. As he himself had said: 'A good general may be beaten, but he should never be surprised.' He ordered the assaults on the Raevski Battery and Semenovskaya to be called off (here he was a little late as Eugène had already done this) and for urgent action to be taken to secure the left flank. Roguet's division of the Young Guard was ordered north up the Kamenka stream to the Kolocha and the Vistula Legion was brought forward to the Kamenka. Napoleon himself rode up to the new Moscow road and stayed there until it became obvious that Platov and Uvarov could achieve nothing.

After about two hours, Barclay sent orders to the two commanders to withdraw as it appeared that nothing was now being gained. Clearly, the Russian high command did not appreciate the breathing space that had just been won or the confusion which had been caused in the enemy's ranks. At about 4 o'clock, Platov and Uvarov, who had been lurking about on the plain, moved off to return to Gorki. When Uvarov reported to Kutuzov on the outcome of the action, the perception in the Russian headquarters of these events was such that Kutuzov replied: 'I know. May God forgive you.'

Eugène and the bulk of his forces had returned to the south bank of the Kolocha some two hours before and were now preparing for the final assault on the Battery.

The Storm of the Grand Battery

This phase of the battle presents us with an example of heavy cavalry doing the job for which it was designed: punching through the weakened centre of an enemy line. It also gives a telling example of why cavalry should not be required to hold ground as opposed to taking it.

At about midday Latour-Maubourg's IV Cavalry Corps was ordered to take post between the captured village of Semenovskaya and the Raevski Battery to plug the gap in the French line caused by the withdrawal of Prince Eugène's IV Corps. Here the cavalry were subjected for some hours to effective shot and shell from the Russian batteries in and around the famous earthwork. The site on which General Lorge's 7th Cavalry Division was placed (east of the junction of the Semenovka and Kamenka streams) was in full view of the guns on this ridge, and only some 600 yards away from them. The order that put them there caused a totally needless waste of lives. It is baffling that no commander dared to point this out and have them moved into cover.

Roth von Schreckenstein, a junior officer in the Saxon

Zastrow Kürassiers in the battle, later published an account which gives us some idea of the desperate situation in which these regiments found themselves:

> 'As men and horses were being shot all the time, the men were fully occupied closing to the centre and telling off in their new files of three; this constant telling off never stopped.
>
> Latour-Maubourg saw that if he could move his corps somewhat to the left, they would benefit from the cover afforded by a gentle dip in the ground. He was loth to have his command carry out a full flank march to the new site in case the Russians should seize the opportunity to attack them as they did so. He thus sent repeated orders to his subordinates, by means of his ADCs, to carry out several minor "shuffles" in order to achieve the same objective.'

Not having his commander's aims explained to him, General Thielemann (one of Lorge's German brigade commanders) became increasingly frustrated by the repeated visitations of these minions with apparently pointless instructions, and eventually, he boiled over. It was the bad luck of one of Latour-Maubourg's ADCs, a young Pole, to come to Thielemann's brigade to deliver the next order for it to move to the left just as this happened. He arrived just as Thielemann was between the Garde du Corps and the Zastrow Kürassiers exchanging his wounded horse for a new mount. Not seeing the general, the ADC delivered the order directly to the regimental commanders. Much to Thielemann's surprise, his brigade began to move off, apparently of its own accord. Spotting the ADC, Thielemann rode up and demanded to know why the order had not been delivered to him. The hapless ADC replied that Thielemann 'had not been at his post'. This was the last straw for the general. Flying into a rage, he drew his sabre and charged straight at the Pole, chasing him all the way back to Latour-Maubourg. Here, he explained to his astounded commander, that he was not one of those who would allow himself to be ordered about by adjutants, much less be insulted by them. Furthermore, if the said ADC showed his face anywhere near him again, he would run him through! It says much for Latour-Maubourg's regard for Thielemann's proven reputation as a brave and competent

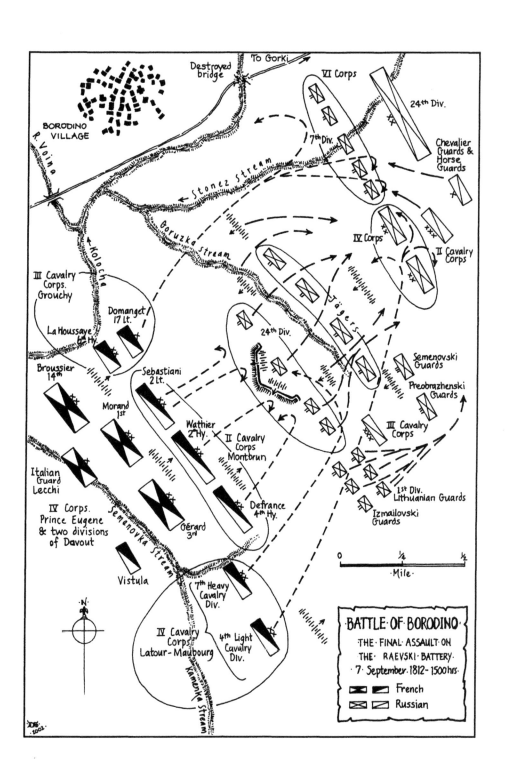

commander, that he did not at once arrest him but calmed him down and took the trouble to explain his aims to him in detail.

Montbrun's II Cavalry Corps was subject to a similar ordeal when it advanced into line north of Lorge's division, just west of the Semenovka stream. The general himself was mortally wounded by a shell splinter as he rode at a walk along the front of his regiments. 'Good shot!' a Prussian trooper heard him murmur as he slid from his horse. Montbrun died at 5 o'clock that evening from the effects of his stomach wound. He was regarded among the Germans in the Grande Armée as being one of the most honourable men in the French Army.

Among Montbrun's regiments was the 1st Prussian Ulans, commanded by Major von Werder. At one point, von Werder's horse collapsed on top of him, killed by an artillery shot. When his men came to drag him out, they found him still with his pipe in his mouth! A shell exploded between the legs of General Subervie (commander of the 16th Light Cavalry Brigade) but left him completely unhurt while two Prussian Ulans beside him (Arnold and Galopy by name) each lost a foot.

A similar fate befell the Bavarian and Saxon cavalry in the 17th Light Cavalry Brigade of Chastel's 3rd Light Cavalry Division, Grouchy's III Cavalry Corps. After they had supported Eugène's second assault on the Raevski Battery they were forced to stand from 10 o'clock in the morning to 3 o'clock in the afternoon in the angle east of the confluence of the Semenovka and Kalotcha streams, enduring the close range fire of the Russian artillery ranged along the northern end of that slight ridge. Caught like rats in a trap, the Saxon Prinz Albrecht Chevauxlégers lost over half their strength in this period to no purpose at all. It was only when Prince Eugène made his third (delayed) assault on the Battery, that this brigade was released from this senseless torture, only to be ripped to pieces by volleys of canister as it breasted the ridge north of the work and received fire from another Russian battery on the next ridge south of Gorki.

When Prince Eugène's troops returned to their position in front of the Battery, they prepared to make the third assault. This time, the infantry were to aim for the north flank of the structure, Montbrun's II Cavalry Corps was also to assault the

The Battle of Borodino

northern side of it, but to strike before the infantry arrived, and the IV Cavalry Corps was to swing around into it from the south.

As Roth von Schreckenstein wrote:

'General Auguste de Caulaincourt [one of Napoleon's ADCs and brother to the French ex-ambassador to the czar] was by now attached to the II Cavalry Corps but only in the capacity of a brigade commander, not at divisional or even corps level as some writers have suggested in the past. This error has been compounded by the King of Naples [that is Marshal Murat], who wrote in his Report (as he related the capture of the heights of Semenovskaya by Latour-Maubourg between 10.30 and 11.30): "Je fis alors passer le General Caulaincourt a la tête du 2. Corps de reserve; a peine fut il de l'autre côté du ravin, que je lui donne l'ordre, de charger sur la gauche tout ce qui se trouve d'ennemis et de tacher d'aborder la grande redoubte qui, nous prenant en flanc, nous faisait boucoup de mal, s'il trouvait l'occasion favourable. Cet ordre fut executé avec autant de célérité que de bravour."[1]

This "occasion favourable" did not occur until much later, after 2 o'clock, when Prince Eugène made his final assault on the battery, probably at the same time as Thielemann received his orders for the same attack. At that point however, the King of Naples was close to the village of Semenovskaya and a long way away from the II Cavalry Corps.

By saying that Caulaincourt was '"at the head" of the II Cavalry Corps, Murat gives the distinct impression that he was actually in command of it. Chambray on the other hand, has Caulaincourt at the head of Watier's division after that general had been wounded.'

The assault was preceded by a barrage of shot and shell from the 170 French and allied guns in the sector which were so

---

1. 'I passed General Caulaincourt at the head of the 2nd Reserve [Cavalry] Corps; he was climbing the other side of the ravine, when I gave the order to charge to the left, just where we had seen the enemy to the side of the Grand Redoubt, which, as we were coming from the flank, was advantageous to us and we had chosen a favourable time. The order was executed with alacrity and bravery.'

arranged as to bring very effective cross-fire to bear on their target. The loss, earlier in the day, of the commander of the Russian Reserve Artillery, General Count Kutaisov, had meant that the great majority of the 300 Russian reserve guns remained idle throughout the battle, and the answering artillery fire was relatively ineffective. By now, the ramparts of the Raevski Battery had been pounded into a shapeless mass, much of which had fallen into the ditch. The allied gunfire was so effective that Colonel Nikitin's battery of horse artillery, which was in action next to the Raevski Battery, lost 93 men and 113 horses in less than an hour. All eight of the reserve horse artillery batteries in this sector of the front were fully employed countering the allied barrage.

Defence of the Raevski Battery was now the responsibility of the gun crews themselves and the VI Russian Corps of General Docturov. This corps consisted of General Likhachev's 24th Division (seven battalions), which were deployed directly behind the battery, in the valley of the Goruzka stream, with Kapsevich's 7th Division to the north. The remnants of the IV and VII Russian Corps were slightly further to the east, on the next ridge, with the II and III Russian Cavalry Corps behind them. Those allied artillery projectiles which missed targets in the front Russian lines, found many a billet in these exposed reserve formations.

It had been the aim that General Raevski's VII Corps was to reorganise behind the reserve cavalry and then advance into line again south of the 24th Division. Despite all the efforts of General Paskevich, however, it was not until nightfall that his 26th Division from VII Corps could be brought forward again.

It was just before 3 o'clock when Eugène's third assault was finally mounted. The infantry divisions of Morand (1st) and Gérard (3rd) formed the first line (Gérard's on the right), with Dommanget's 17th Light Cavalry Brigade on the left wing, up against the Kolocha. In the second line were Lecchi's Italian Guard (on the right) and Broussier's 14th Division with La Houssaye's 6th Cavalry Division on the left. Claparède's Vistula Legion had advanced up from Shevardino to take post on the right of Lecchi's division.

The posting of the cavalry on the left wing was, of course, in

The Battle of Borodino

line with classic military teaching. In this situation however, it might have been more appropriate to make an exception to the rule. Bearing in mind that the cavalry was hemmed in on their left by the Kolocha, which was here a definite obstacle as it ran through its deeply-cut channel, the chances for their successful deployment were as good as nil. They could equally well have been deployed behind the infantry and would still have been available to exploit any tactical opportunities which might have arisen in the course of the action. South of Gérard's division was what was now Sebastiani's II Cavalry Corps then Latour-Maubourg's IV Cavalry Corps.

Meanwhile, Lorge's 7th Cuirassier Division was suffering the continued effects of the Russian artillery bombardment with great stoicism. In fact, now that it had moved a little further to the left, into a slight dip in the ground, its casualty rate had decreased somewhat. This relief was amplified by the fact that many of the shells which landed among it now had had their fuzes set too long. Instead of bursting in the air just above their target in order to create maximum havoc, most buried themselves in the ground so that the fuzes were snuffed out and the shells did not detonate. The situation became so relaxed that General Thielemann ordered the men to break out their weeks-old rations of hard tack biscuits and take a frugal snack.

Just before 3 o'clock Eugène's infantry moved off to the assault. Shortly afterwards, one of Napoleon's ADCs galloped up to the impatient Thielemann and delivered to him the long-awaited message: 'On behalf of the Emperor, I bring you the order to attack!' This same order had also been delivered to the II Cavalry Corps. Caulaincourt led Watier's 2nd Cuirassier Division forward against the north flank of the Raevski Battery. Sebastiani's 2nd Light Cavalry and Defrance's 4th Cuirassier Divisions advanced directly on the front of the work, while Lorge's 7th Cuirassier Division advanced on its southern flank. Barclay saw the movements in the enemy ranks opposite him and sent orders for his 1st Cuirassier Division to advance from its position behind the Lifeguard Division to counter the attack. Imagine his astonishment and rage when his ADC reported that the division was no longer there but had been sent by someone to the far southern wing earlier in the day!

That such an arbitrary move could have been carried out without Barclay having been consulted or informed speaks volumes on the practices in the Russian General Staff. As Barclay said:

> 'When I saw that the enemy were about to launch a violent assault, I at once sent for the 1st Cuirassier Division, supposing it to be still in the position which I had assigned to it, for I intended to keep it for a decisive blow. Unfortunately, someone – I don't know who – had moved it to the extreme left flank. All my adjutant could do was to collect just the two regiments of Lifeguard Cuirassiers and bring them to me by companies as quickly as he could.'

By the time these two regiments had assembled, the three enemy corps were well into their advance and it was clear to Barclay that it would be suicide to commit his men against such odds.

What actually happened in this massive allied cavalry assault has been obscured by dubious, partisan French reports, generated at Napoleon's instigation and scarcely questioned since then. Which regiment was it that was first into the Battery? Which regiment actually took and held it? As we shall see, Napoleon's official rumour mill worked at top speed and with its customary efficiency to ensure that posterity would be forced to accept his version of events. In the years following the battle, however, still, small voices could be heard, by those willing to listen, which must lead us to treat this Napoleonic Bulletin with the circumspection now accorded to so much else that he wrote, or commanded to be written.

Meerheim, an officer in the Saxon Garde du Corps, described the final assault on the Raevski Battery as follows:

> 'The redoubt lay at the top of a steep slope rather like the one which we had had to climb to get into Semenovskaya. [This I cannot understand; the slope from the west up to the site of the redoubt is level and gentle.] It was covered by a fairly wide ditch but built, happily, only of loose earth, without palisades and obviously constructed in a hurry.
>
> On the side facing us there was a deep, narrow valley [not today] like a second ditch which we also had to cross before we could storm the actual crest of the ridge.

The Battle of Borodino

> In the Battery itself were maybe 12 or more guns; the remaining space was filled with infantry. The ditch and the "ravine" were also filled with infantry.
>
> Behind the battery were several, fairly strong squares of infantry relatively close together and ranged along the far side of the valley east of the ridge on which the Battery stood.
>
> We saw (or rather felt) the presence of a strong force of artillery also on that ridge.
>
> Apart from this, we could see several lines of infantry and cavalry in reserve further back. In the dip towards Borodino [along the Goruzka stream] were more masses of infantry and cavalry which had previously been concealed from our view.
>
> We charged at the "ravine" and ditch, the horses clearing the bristling fences of bayonets as they would have chevaux de fris [sic]. The combat was frightful! Men and horses hit by gunshots collapsed into the ditches and thrashed around among the dead and dying, each trying to kill the enemy with their weapons, their bare hands or even their teeth. To add to this horror, the succeeding ranks of assaulting cavalry trampled over the writhing mass as they drove on to their next targets – the infantry squares – who greeted them with well-aimed volleys... '

Meerheim's description then speaks of some Russian infantry in huts in the valley to the east of the battery firing out of the windows at the attackers but I can find no trace of a similar account of such constructions and the nearest village east of the battery was Kniaskovo, over a mile away. Meerheim then continues:

> 'Despite all the perils and obstacles we were unstoppable and burst over and into the Battery, inspired by the examples of our commanders, Generals Latour, Thielemann and our brigade adjutant von Minkwitz. The interior of the Battery was an indescribable mess of infantry and cavalry all intent on killing one another. The garrison of the place fought to the last.'

Napoleon (and many historians since) attributed the final capture of Raevski's Battery to French cuirassiers under the leadership of Caulaincourt. This on-the-spot, eyewitness account, together with the carefully conducted investigation of Roth von Schreckenstein contained in his book *Die Kavallerie in der Schlacht an der Moskwa*, should finally lay this ghost. As von Schreckenstein wrote:

> 'If I return again to the Report of the King of Naples of 9 September, concerning the participation of the Reserve Cavalry in the battle, it is because this report contains a number of inaccuracies that have been repeated in subsequent documents including the Bulletin by the Marquis de Chambray of 10th September.
>
> ... I have already shown how completely wrongly Chambray describes Latour-Maubourg's attack on Semenovskaya village. Chambray is equally unclear... concerning the capture of the Raevski Battery where he has Caulaincourt enter the work by wheeling to the left, having previously charged a line of the enemy.
>
> This is what he [Chambray] wrote: "Eugène ordered the divisions of Broussier, Morand and Gérard to cease firing and to storm [the Battery]. At the same time, Caulaincourt, at the head of Watier's division, overthrew the line of the enemy that was opposite to him, then, wheeling left, charged through the troops close behind the redoubt then, turning back towards the work, entered it from the rear. Eugène stormed over the parapet from the front at this same instant; all who defended themselves were cut down. Twenty-one guns fell into French hands. Watier resumed his position on Eugène's right flank. Caulaincourt had been fatally wounded in the redoubt itself. It was now three o'clock..."
>
> According to this, it seems that the Marquis de Chambray was ignorant of the fact that it was only Defrance's division which assaulted on the right of the redoubt and that Watier's division attacked it from the direction of Borodino [i.e. from the northern side], whilst the Saxon cavalry penetrated into it from the direction of Semenovskaya [the south].
>
> This writer [Chambray] relates nothing of the active participation of Latour-Maubourg's corps here because there was nothing about it in the Bulletin and nothing in the Report.
>
> I feel that it is quite possible that Caulaincourt was wounded near the Battery, but I feel that it is quite

The Battle of Borodino 139

unacceptable for the King of Naples to bury the victor's laurels with him.

I further dispute the claim that he overthrew a line of enemy infantry behind the redoubt. It is incredible that the King of Naples seems to have forgotten that Montbrun was in command of all the II Cavalry Corps at the redoubt and that he was so close to the Battery two hours before the final storm that he lost his life.

It is also astounding that the King brings the capture of the heights of Semenovskaya (which was completed by 12 o'clock) directly into context with the final assault on the Raevski Battery which, as is well known, took place at about 3 o'clock. It is as if the impression is meant to be given that nothing happened on the battlefield for three hours; that Latour-Maubourg was completely inactive.

Or perhaps sections of the King's Report were removed at Napoleon's instructions in order to glorify Caulaincourt.

Latour-Maubourg was the senior allied cavalry commander in the vicinity of the Battery [at the time of its capture]; the charge into the Battery was only the preliminary to its final capture which was utterly dependent upon the defeat of the Russian infantry arrayed behind it. There is no doubt that it was Latour-Maubourg's cavalry corps, supported by Defrance's division which carried out this difficult task. I am in a position to confirm that he gave all the orders on the right hand side of the Battery and that a considerable time elapsed before the Viceroy's infantry came up.

According to an eye-witness who related the following to us the next morning, Napoleon was standing by Berthier [at Shevardino] when the latter, squinting through his telescope, said "The redoubt is taken, the Saxon cuirassiers are in it!" Napoleon took the telescope, peered through it and said: "You are wrong, they are dressed in blue; they are my cuirassiers." What Napoleon may have seen through his telescope, from his distant vantage point at Shevardino through all the smoke and confusion, was the 14th Polish Cuirassier Regiment which at this point formed the third rank of Thielemann's Saxon brigade as it followed the two Saxon regiments into the Battery. The uniform of the Polish cuirassiers was almost identical to that of their French comrades. Or they may have been the Westphalian Kürassiers [of Lepel's brigade, 7th Division] who followed us.'

In any case, Napoleon's regard for the truth was such that he would have said whatever he wished to have posterity to believe, knowing full well that none of those present would have dared to contradict him. Roth von Schreckenstein continues:

> 'One thing is certain, in his 18th Bulletin Napoleon attributed the capture of the battery to Caulaincourt and mentioned the IV Cavalry Corps only in connection with the capture of Semenovskaya. Latour-Maubourg was extremely angry at this misinterpretation of the true events.'

As von Schreckenstein stated, the capture of the Battery itself was nothing without the defeat of all the Russian infantry in the valley of the Goruzka stream directly behind it. All eye-witness accounts of the final storming of the battery speak of being surprised at the presence of strong Russian infantry forces 'concealed in the steep-sided ravine' directly to the east of the work. The heat of battle may have distorted individual memories of the sequence of events, their timing and of the terrain on which they occurred, but it is inexplicable that so many should agree that there was such a well defined geographical feature there. There is a valley it is true, but it is only marginally more steeply sloped than the western side of the ridge on which the Battery was built. The valley clearly did, however, offer excellent cover from enemy view and fire (except howitzer shells). As to how far this may be interpreted as an intentional use of the terrain by the Russians to conceal and protect their forces must remain up to the reader to judge; at least no Russian commander claims credit for having used the ground for this purpose, and several bewail the fact that it was not done, thus contributing to the very heavy casualties that they suffered.

Whoever was first into the battery in this final assault, it is clear that the cavalry had reached it before the Viceroy's infantry and that the II, III, and IV Cavalry Corps were all involved. The cavalry, of course, could take the ground, but it required infantry to hold it for any length of time. When the massed allied cavalry rolled forward to assault the insignificant, battered

The Battle of Borodino

earthwork, the Russian infantry flanking it on the crest of the ridge formed square as the command, 'Prepare to receive cavalry!' rang out. This meant, of course, that the allied artillery, firing as they were over the heads of their own cavalry, could cause much more havoc in the densely packed Russian regiments than when they were in three-deep line.

The Russian historian Bogdanovich described the final charge as follows:

> 'Watier's division crossed the Semenovka stream at the mouth of the Kamenka; passed left of the Raevski Redoubt and hit part of the VI Corps in the Goruzka valley. The 5th Cuirassiers turned right and went over the ditch and rampart into the redoubt but were forced out again by infantry fire. Caulaincourt was fatally wounded by a musket ball to the throat.
>
> General Bachmetiev I (commander, 23rd Division) lost a leg; Generals Bachmetiev II and Alexopol were badly wounded, General Ostermann was slightly wounded.
>
> Defrance's division should have passed left of the redoubt together with [Watier's] cuirassiers, but was late coming up. Meanwhile, Latour-Maubourg's Cavalry Corps came around the left side of the redoubt. Rozniecki's Ulan [4th Polish Light Cavalry] division, in two ranks, formed the right wing of the assault; the cuirassiers were the left wing; in the centre were the horse artillery batteries.
>
> The Saxon Garde du Corps went for the Redoubt, the Saxon Kürassier Regiment Zastrow and the Polish Cuirassier Regiment Malachowski and the Westphalian cuirassier brigade charged the Infantry Regiments Pernov, Kexholm and the 33rd Jäger who were to the left in the valley. Our Russian infantry fired a volley at 60 paces; the enemy cavalry fled.
>
> General Thielemann and the Saxon Garde du Corps went over the ditch and the rampart into the redoubt. The general officer in command of the defence of the redoubt, General Likhachev, was in the redoubt. He was sick and also wounded, and was captured when the place was taken.'

Likhachev's presentation to Napoleon following his capture is the subject of a dramatic oil painting, set in the usual 'Alpine' landscape and lit by the heavenly spotlights so beloved of those who were commissioned to immortalise the Emperor's conquests. It seems that Napoleon wished to return to Likhachev his sword, which had been taken from him, in recognition of his gallant defence of the Battery. Beckoning to an aide, who had accompanied Likhachev into his presence, Napoleon took the extra sword that the man was holding and proffered it to the captured general. Not recognising his own weapon, Likhachev refused to accept it. Apparently, linguistic difficulties prevented a clarification of the situation. In the end Napoleon, becoming impatient at the embarrassing situation in which he found himself, simply muttered: 'Take this idiot away!'

Unbelievable as it may seem, in the midst of the bloodbath and mayhem of the final struggle for the Battery, the Russian artillery managed to bring four of the 12 pounder pieces out and to bring them to the rear. Bogdanovich continues:

> 'At this point, Prince Eugène reached the redoubt with his infantry (the divisions of Morand, Gérard and Broussier) and put the 9th Line [of the 14th Division] into it as garrison.
>
> Barclay de Tolly ordered the 24th Division to retake the redoubt from the valley to the east of it. Malachowski's cuirassiers charged the advancing Russians in their left flank. The charge was beaten off but Barclay now ordered a withdrawal behind the Gorizi [sic] stream.
>
> Enemy cavalry now prepared to charge the squares of the IV and VI Corps. A Polish Ulan regiment attacked the Russian foot artillery battery in which General Kostenezki, commander of the artillery of the VI Corps found himself. Kostenezki was a tall, strongly-built man, he seized a rammer and led a counter-attack by the gunners to drive the Polish cavalry off.
>
> Most of Barclay's cavalry had been detached to our left wing earlier in the day, so he now had to conduct his defence with only infantry and artillery, but – at Barclay's orders – the Chevalier Guard and the Horse Guards under Major-General Shevich came up from the reserve and took post behind a small hill before Kniaskovo.
>
> Prince Eugène's IV Corps, Defrance's 4th Cuirassier Division, Chastel's 3rd Light Cavalry Division and the 7th

The Battle of Borodino 143

Dragoons [of La Houssaye's 6th Cavalry Division] now attacked Kapsevich's 7th Division in front and rear [in its location north of the Battery]. Defrance's Carabiniers rode down a square of the 19th Jäger Regiment.

At this point, the 2nd division [section] of the 2nd Guards Horse Artillery Battery, of Lieutenant Baron Korf, came up from the Reserve to the right hand side of the Chevalier Guard. Seeing the defeat of the Jägers, Korf – on his own initiative – advanced to within 300 yards of the enemy, unlimbered, opened up with canister and drove the enemy back.

Only Chastel's Light Cavalry Brigade remained on the field and these now charged at Korf's battery. Korf appealed at once to the commander of the 1st Troop, right wing squadron of the Chevalier Guards [of the 1st Cuirassier Division]: "Bashmakov! Save my guns!" The Chevalier Guard charged and drove off Chastel's brigade, saving the battery.

Now Barclay came up and ordered the Chevalier Guard to undertake another charge. Colonel Loewenwold [the commander] formed his regiment by squadrons, advanced through the intervals of the squares of the 19th and 40th Jägers and charged at the enemy cavalry of Thielemann's brigade. These were formed up with the Saxon Garde du Corps and the Kürassier Regiment von Zastrow in the first rank, with two Polish lancer regiments in the second.

As the Russians closed with the enemy, they were received with a discharge of canister; Loewenwold fell, killed by a canister ball to the head. Colonel Levachov took over command at once but the momentum of the charge had been broken.'

Apparently, this incident was only one of many such clashes between the Saxons and the Chevalier Guard which took place in this closing phase of the battle to the south-east of the Battery; Roth von Schreckenstein tells us that, eventually, the Russians had the best of it and the Saxons retired. Bogdanovich then continued:

'Renewed cavalry charges by the Chevalier Guard and the Horse Guards stabilised the situation [this was to the south-east of the Battery when the Izmailovski and Lithuanian Guards of the 11th Infantry Division – in six battalion squares – withdrew east across the valley to their new position on the ridge south of Gorki].

Enemy pressure on Barclay de Tolly increased; Barclay ordered General Adjutant Korf's II Cavalry Corps to advance [on his northern flank] to help him hold on. Korf sent Major-General Panchulidsev II with the Isum Hussars and the Polish Ulan Regiment who were sent against Watier's cuirassiers and Defrance's Carabiniers. Their charge was beaten off but they rallied quickly.

Korf now sent Colonel Sass's Pskov Dragoons, with the Moscow Dragoons in support, to assault the right hand side [north] of the Raevski Battery. On the way there, Sass saw enemy infantry and cavalry threatening the the right flank of the Isum Hussars and the Polish Ulans. He at once charged the enemy cavalry [it may have been the 7th Dragoons] and – having put them to flight – continued on towards his original target. He was now involved in a clash with enemy cavalry in which the French were eventually beaten back.

The Dragoon Regiment Pskov and the 1st division, 2nd Guards Horse Artillery Battery, under Colonel Kosen, now assaulted the enemy's left wing battalion, but a 12-gun French battery caused them heavy loss and the attack failed.

The III Cavalry Corps now came up with the Soum and Mariupol Hussars, the Orenburg, Siberian and Irkutsk Dragoons. Supported by part of the II Cavalry Corps, they charged the enemy cavalry masses. There followed a violent cavalry mêlée. Barclay de Tolly and his staff were right in the middle of it. One of Barclay's adjutants, Count Lamsdorf, was killed by an enemy pistol shot.

On the French side, Generals Chastel, Huard and Gérard were killed; Grouchy and Domanget [sic] were wounded.

Prince Eugène's IV Corps now had control of the Raevski Battery and the ridge on which it stood.

It was now 4 o'clock. The Russian Army was in a line, 1,200 paces east of the Raevski Battery, east of the Gorizki valley. The right wing was at Gorki, the left wing was a cannon shot east of Semenovskaya, on the edge of the woods. The fighting now died away, except on the southern flank.'

The Battle of Borodino

Thus ended this epic and costly struggle. Both sides had given more than their all. The cost in blood had been tremendous; the result – a draw.

Incredible as it may appear, there were troops on both sides who had not been engaged in this battle. On the French side Napoleon refused repeatedly to commit the Imperial Guard to the battle despite the urgent entreaties of his marshals. In addition there was Claparède's Legion of the Vistula, which had been brought forward in the afternoon but not actually thrown into the combat. Delzons's 13th Division was also reasonably intact. This gave a total of about 25,000 men. The opinion of posterity has come down on Napoleon's side in this decision. As he said, he was not going to risk his last reserve so far from home unless it was absolutely necessary.

On the Russian side, the 4th, 30th, 34th and 48th Jägers (on the extreme right wing behind Gorki) had not been engaged all day. The Preobrazhenski and Semenovski Guards Infantry Regiments had not been committed, and there were 15,000 militia, Uvarov's I Cavalry Corps and Platov's Cossacks which had scarcely been engaged. The Artillery Reserve of the 1st Army had also been largely unemployed, but had suffered considerable damage and casualties as it stood idly on the ridge behind Kniaskovo. In all, there were some 28,000 relatively fresh troops still available, due more to luck and mismanagement rather than skill.

Sources

Bogdanovich, M.I., *Geschichte des Feldzuges von 1812*, Leipzig, 1863

Chandler, David, *The Campaigns of Napoleon*, London, 1966

Esposito, V.J., and Elting, J.R., *An Atlas of the Napoleonic Wars*, London, 1999

Martinien, A., *Tableaux pars Corps et pars Batailles des Officiers Tués et Blessés Pendant les Guerres de l'Empire 1805–1815*, Paris, 1890

Roth von Schreckenstein, Freiherr Generalleutnant, *Die Kavallerie in der Schlacht an der Moskwa 7. September 1812*, Münster, 1855

Six, Georges, *Dictionnaire Biographique des Generaux & Amiraux Français de la Revolution et de l'Empire (1792–1814)*, Paris, 1934

Chapter 8
The Battle of the Beresina Crossing
28 November 1812

One of the most desperate struggles of the entire Napoleonic Wars was that which unfolded on the banks of a minor Russian river in late November 1812. Here, the French emperor, having failed utterly to bend the czar and his people to his will, was leading the few survivors of his once-massive army back from Moscow, keeping just one step ahead of the main Russian army under Kutuzov while other Russian forces were also converging on the invaders. It was a race to see if the French could unite their formations and cross the Beresina, before they were crushed by their pursuers.

In this action, we will see how two German cavalry regiments were sacrificed to save the Grande Armée. In November 1812 Marshal Victor's IX Corps of the Grande Armée consisted of the 12th, 26th and 28th Infantry Divisions and two brigades of light cavalry. The 12th Division was made up of nominally French (but included some ex-Dutch) regiments; the 26th of Germans from Baden and Berg, and the 28th was a Polish division. The cavalry were General Fournier's 30th Light Cavalry Brigade, 2nd Lancers of Berg and the Guard Chevauxlégers of Hesse-Darmstadt, and General Delaitre's 31st Light Cavalry Brigade which had the Saxon Chevauxlégers Regiment Prinz Johann and the Baden Hussars.

The IX Corps had initially been charged with duties on

The Battle of the Beresina Crossing

Napoleon's lines of communication through East Prussia, but in August it had been ordered forward into Russia. The corps crossed the River Niemen at Kovno on 30 August, with a strength of 33,500 men, and marched on Vilna. At this point, the state of the cavalry regiments of the corps was not too bad, despite the lack of proper fodder, including the all-important oats. On 14 August the Hessians recorded having lost only 11 horses since the start of the campaign. This happy situation was soon to deteriorate rapidly as they advanced into the desert that the Russians and the Grande Armée had created in the rear zones of the war.

By 29 September IX Corps had reached Smolensk. The regiments were forced to forage in order to exist; the only items supplied from the huge magazines which had been stockpiled in the city were salt and brandy. Roving Cossack bands were constantly harrying the foraging parties. On 18 October the Grande Armée left Moscow and began to withdraw westwards towards safety, a safety that pitifully few of them would ever live to see. Now Victor's corps was ordered back west from Smolensk to Vitebsk to block the advance of Wittgenstein's I Corps, which was pushing the remnants of the II and VI Corps of the Grande Armée southwards before it. Victor and his men were to hold the position there until the main body of the Grand Armée had crossed the River Dnieper at Orsha.

On 29 October the IX Corps (strength now down to about 16,000 men) joined up with Gouvion St. Cyr's II Corps at Chashniki, and for two days they and Wittgenstein's Russians faced off against one another, before the French forces withdrew south to the main road east of the River Beresina. For the next two weeks the IX Corps stayed in Senno and then moved off south-west to join up with the main body of the army, retreating from Moscow. On 25 November, the men of the IX Corps met the returning, starving, ragged scarecrows of the Grande Armée at Loshnitza. They were shattered and deeply demoralised at the spectacle.

At this point, the situation of Napoleon and the remnants of his army was extremely critical. Wittgenstein's corps and Admiral Chichagov's Army of the Moldau were close to joining up to to attack. Some days before, when the cold snap was most

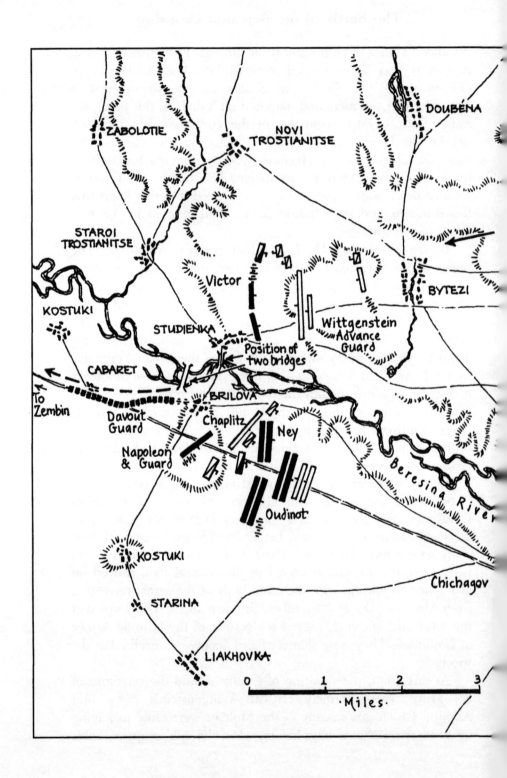

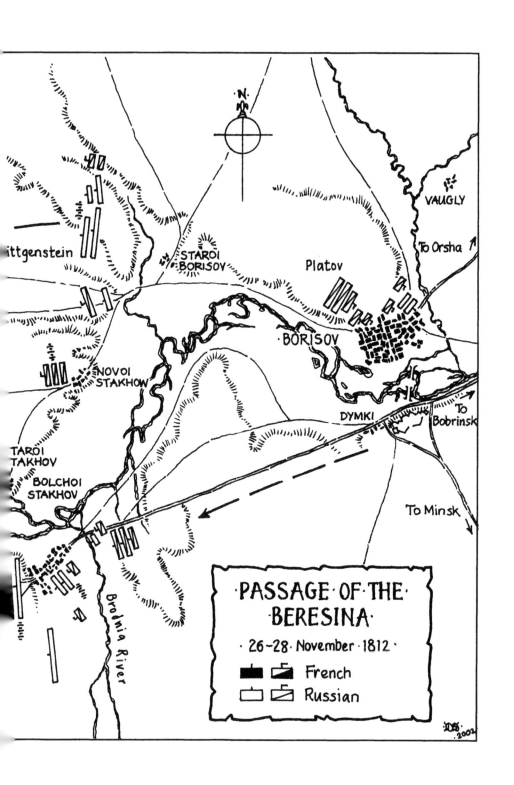

severe, Napoleon had ordered his bridging train to be burned in Orsha. Since all the rivers were frozen he thought there would be no more need to build bridges to cross them. But, by a cruel twist of fate, there was a dramatic rise in the temperature just after this had been done. The rivers were now flowing again and presented major obstacles to anyone attempting to cross them if there were no bridges nor boats available. The men of the Grande Armée was still to the east of the Beresina; it had to be crossed if they were to survive.

A bridge over the river had existed at Borisov but Chichagov's advanced guard under General Lambert had seized this from Dombrowski's 17th Division on 21 November and broken the bridge. The Russians still held the bridgehead on the western bank despite furious French attempts to retake the place. Two other bridges had been located some ten miles to the north, between Kosinki and Veselovo on the road from Vitebsk to Vilna, but these too had already been destroyed. Between these two sites, Napoleon and his army were trapped up against the slowly flowing river – with no bridging train.

Luckily for Napoleon, the true state of his army was not known to Kutuzov, who was following slowly, some distance behind him. Even more luckily for the Emperor, his Commander of Pontonniers, General Count Jean-Baptiste Eblé, had disobeyed his orders to destroy all the bridging equipment. He had kept two field forges, two wagons of coal and six wagons loaded with beams, nails, clamps and tools. Also, each pontonnier still carried his tools and a quantity of nails. In the coming days, these, and the dedication of the French pontonniers, would provide the golden escape route for Napoleon and his men.

General Count Piotr Pahlen, of the Army of the Moldau, had learned from French prisoners on 22 November, that Napoleon was close by and would try to cross the Beresina at Borisov very soon. He sent news of this to Chichagov and to Kutuzov. Neither believed him. Neither moved in for the inevitable kill with any semblance of urgency.

Meanwhile, the desperation of the situation had galvanised Napoleon's great mind into action. Learning that certain bridging tools and materials still existed, he planned to build two bridges over the river between the villages of Studienka on the

The Battle of the Beresina Crossing

east side and Brilova on the west. The houses in the former village would provide the materials needed. As Napoleon said: 'I have been emperor long enough; it is time now to be a general.' The Grande Armée owed its salvation to the foresight of General Eblé, and to the dedication of his 150 pontonniers, many of whom died in the icy waters of the Beresina as they built these two bridges across it.

To mask his plans from the immediate enemy on both sides of the river, Napoleon staged a diversion. On 26 November he sent General Junot, with the remnants of his VIII (Westphalian) Corps, to the area of Borisov to make loud and obvious bridging attempts. Local peasants were leaked the information that this was indeed the Emperor's chosen bridging point. This information was quickly passed on to Wittgenstein and was regarded as credible. The main Russian forces thus remained in this area and only light Cossack patrols were left to observe the Studienka site. Some idea of the state of Napoleon's once-great army may be gleaned from the fact that the entire VIII Westphalian Corps spent the night in one house in Borisov during this operation.

By the 26th Napoleon had had some rafts built and was able to send a detachment of cavalry wading through a shallow spot in the Beresina at Studienka, each cavalryman with a voltigeur behind his saddle. Some 300 infantry ferried themselves over the river on the rafts and, together, these men drove off the Cossack patrols. The construction of the two bridges began with feverish haste at 8 o'clock in the morning of the 26th. The river was about 100 yards wide at this point and the bridges were sited about 200 yards apart. The northern one was for guns and vehicles, the southern for infantry and cavalry. The infantry bridge was finished at 1 o'clock in the afternoon; that for the artillery was ready by 4 o'clock. The two structures were fragile; their surfaces often dipped below the level of the water under their loads and each had to be repaired on several occasions.

Oudinot's II Corps (5,600 infantry, 1,400 cavalry and two guns) was rushed across and took post on the west bank, facing south. Their mission was to fend off any Russian attack and to keep the vital withdrawal route to Zembin open. Soon, Chaplitz's 5,000-strong advanced guard of Chichagov's army came up from

Borisov, but Oudinot's men held firm, supported by the fire of a 50-gun battery which Napoleon had set up on some high ground, east of Studienka. As fast as possible, Ney's III Corps and the Imperial Guard doubled over the river to join Oudinot. The Russian assault was beaten back; the way to the west was now open.

By 27 November the IX Corps had reached Borisov on the River Beresina; its strength had now dwindled to about 7,500 men. However, later that same day, General Partouneaux's 12th Division (the strongest of the IX Corps), together with the Saxon Chevauxlégers Regiment Prinz Johann and the 2nd Lancers of Berg were surprised and captured by Wittgenstein's corps some 2½ miles west of Borisov while the corps was moving towards Studienka.

By midnight on 27–28 November most of the formed units of the Grande Armée had crossed the Beresina using the two flimsy bridges. Only then was the great mass of stragglers and fugitives allowed to cross.

We will pass over the grim drama of the actual crossing, with its overpowering human misery, and concentrate on the tactical events. Suffice it to say, that it was recorded by several eyewitnesses, that, although the bridges were packed during the daylight hours, they were deserted at night. The masses of unarmed refugees on the eastern bank, who were to fall into Russian hands, failed to take advantage of their chances to cross the river to relative ease and safety.

Miraculously, at this point, a Baden officer, Lieutenant Hammes, arrived at his brigade's location with a convoy of 41 wagons of shoes, clothing and food, all the way from their homeland, over a thousand miles away! The items were issued out quickly and the morale of the brigade rocketed. It was now 1,240 men strong. That a lowly German lieutenant succeeded in transporting such a large convoy of vital supplies safely to his comrades, halfway across a continent, when Napoleon's genius had been thwarted in the first weeks of this fateful campaign, is almost incredible.

Victor's IX Corps was now given the task of mounting the rearguard action to protect the bridges and cover the retreat of the army on 28 November. His strength was now down to 5,000

The Battle of the Beresina Crossing

infantry, 350 cavalry and 14 guns. The final stand of the corps was made by General Girard's 28th (Polish) Infantry Division, the Baden and Berg infantry brigades, and Fournier's 30th Light Cavalry Brigade, with 14 guns.

In the early hours of 28 November, the remnants of the IX Corps recrossed the bridges to the eastern bank. The commander of the 26th Division, General Dändels, excused his absence from the impending action to General Wilhelm Graf Hochberg of the Baden brigade, explaining that he had no horse and could not command without one. The German general at once gave him the horse of Feldjäger Schütz. Scarcely had Dändels mounted it, than he asked Graf Hochberg to assume command 'if he should not be there'. Shortly after this, Hochberg was informed by Dändels's chief of staff that 'the general had fallen in the water' and, some time later, the chief of staff himself was wounded in the finger and left for the rear. Neither he, nor his commander, was seen again during this action.

During the forenoon of 28 November the IX Corps was assaulted by Wittgenstein's advanced guard under General Vlastov (8,000 men and 24 guns). The infantry of the IX Corps extended from the river south of Studienka in a crescent around the place; but there was no natural feature on which to anchor the left flank so Fournier's cavalry (Baden Hussars and Hessian Guard Chevauxlégers) were posted there.

As the Russian pressure mounted, Victor ordered the Baden infantry to advance and drive the enemy back. This attempt was repulsed with heavy loss. They tried again; again without success. In a final effort to stop the Russian advance, Fournier's cavalry was thrown in. The regimental history of the Baden Hussars recounts events as follows:

> 'Colonel von Laroche charged the Russian infantry, which had formed square and had their artillery with them. He also commanded the Hessian Chevauxlégers and these followed his regiment at a trot. The battalion of the 34th Russian Infantry Regiment [sic][1] fired a volley only when the hussars were very close. Despite this, the square was broken, the

---

1. This should be 'Jäger Regiment' since Russian infantry regiments bore titles, not numbers. The 34th Jägers were originally in the 4th Infantry Division, II Corps, of the First Army of the West.

infantry either cut down or captured. After the hussars had handed over 500 prisoners to the chevauxlégers, they went after a string of enemy infantry in extended skirmishing order. They then fell on the Nizhov and Voronezh Infantry Regiments and the Depot Battalion of the Pavlov Grenadier Regiment. Soon after this, two squadrons of enemy cuirassiers [a provisional regiment made up of depot troops] appeared; Laroche gathered in his hussars and charged them.

In the wild mêlée, Sergeant-Major Martin Springer of the hussars, although wounded already by sabre cuts and bullets, went to the aid of Colonel von Laroche and cut down his opponents, thus freeing the colonel.

As the Germans were fighting the cuirassiers, they were taken in flank by a Russian hussar regiment,[2] which sealed their fate.'

Graf Hochberg's official report gave his version of events as follows:

'The hussar regiment was almost destroyed in this honourable combat. Only 50 horses out of 350 returned over the Beresina. The brave chevauxlégers shared the same fate. It was a stroke of destiny that these two fine regiments were able to close their battlefield careers with such a fine action which saved the lives of so many of their comrades by their sacrifice, in this campaign where the deprived cavalry, in the bitter climate, were facing their end.'

The fight went on for five hours until darkness fell, after which the Russians ceased their assaults. The cost had been heavy. The Baden Hussars and the Hessians had effectively been destroyed. Only 50 hussars were still present. The Baden infantry brigade now consisted of 28 officers and 900 men. The Berg infantry brigade had only Colonel Genty and 60 men. The 28th Polish Division counted just 300 men. Of the Saxon Prinz Johann Chevauxléger Regiment only 109 men remained. Marshal Victor and Generals Damas, Girard and Fournier were wounded. But they had bought the time required of them. The

---

2. The only Russian hussar regiment recorded as being in action at the Beresina on this day was the Alexandria Regiment. This regiment was in General Tormassov's Third Army of the West, attached to the 15th Division.

The Battle of the Beresina Crossing

brave Sergeant-Major Springer was rewarded with a commission and retired as a lieutenant of artillery in 1829. Colonel von Laroche, badly wounded, was saved by some men of his regiment and returned safely to Baden.

Graf Hochberg, as the only general officer left fit for duty, took command of the 'IX Corps'. At midnight, the corps began to withdraw over the river. In the dreadful scramble to get back over the bridges to safety, Colonel von Dalwigk, commander of the Hessian Guard Chevauxlégers, became separated from his men and was pushed into the icy river by the crush. A Hessian driver recognised the colonel by his white greatcoat and pulled him out. Lieutenant Lippert, of the same regiment, was unable to reach the bridge, so he, and one of his troopers, swam their horses across.

Graf Hochberg was one of the last to cross on the morning of 29 November. The bridges were then fired. About 10,000 unarmed refugees on the eastern bank, men, women and children, fell into Russian hands to say nothing of a vast collection of booty, including the imperial treasure chests. The surviving German cavalry of IX Corps was detailed for orderly duties in Victor's headquarters as of that day, as they were too weak to be of further tactical use. The famous 29th Bulletin which Napoleon composed a few days later contained not one word of praise for the efforts of the Germans of IX Corps. It did, however, mention General Fournier, but he had been wounded early in the action and had left the field.

In mid-December 1812 the IX Corps returned through Kovno 3,500 strong; in four months it had lost almost 90 per cent of its strength. The good General Dändels survived to become governor of the fortress of Modlin, where he capitulated on 25 December 1812. In the Waterloo campaign he served in the Anglo-Dutch army.

Thus ended one of the most tragic and dramatic rearguard actions in military history. The survivors of the Grande Armée were now allowed to drag themselves out of Russia with little further interference. On the Russian side, once full realisation dawned of the chance that had been missed, recriminations began to fly. The upshot was that Admiral Chichagov was blamed for the fiasco. It should be remembered, however, that

Chichagov was not the only Russian commander to ignore Count Pahlen's information. Kutuzov's tardy pursuit of Napoleon in 1812 has been a perennial subject for discussion ever since.

Sources

Chandler, David, *The Campaigns of Napoleon*, London, 1966

Esposito, V.J., and Elting, J.R., *An Atlas of the Napoleonic Wars*, London, 1999

Martinien, A., *Tableaux pars Corps et pars Batailles des Officiers Tués et Blessés Pendant les Guerres de l'Empire 1805–1815*, Paris, 1890

Pivka, Otto von, *Armies of 1812*, Cambridge, 1977

Six, Georges, *Dictionnaire Biographique des Generaux & Amiraux Français de la Revolution et de l'Empire (1792–1814)*, Paris, 1934

Smith, Digby, *Borodino*, 1998

CHAPTER 9
The Clash at Haynau
26 MAY 1813

This little town is now known as Chojnów; in 1813 it was in the Prussian province of Silesia but it is now in Poland, between Boleslawiec (Bunzlau in German) in the west and the town of Legnica (Liegnitz), and the Rivers Katzbach and Oder to the east. It is 28 miles south of the fortress of Głogów (Glogau) on the latter river. The German names as of 1813 will be used from here on to avoid confusing the reader.

In the turbulent, uncertain days of early 1813, Prussia had seized the chance offered by Napoleon's catastrophic defeat in Russia in the previous year to unite with the Russians to throw off the hated French yoke. This alliance against the Emperor had been forced onto the cautious, indecisive Prussian King, Friedrich Wilhelm III, by the commander of the Prussian corps in Russia in 1812, General-Leutnant David Ludwig von Yorck. On 30 December 1812 Yorck had concluded the Convention of Tauroggen with General von Diebitsch, a Prussian in Russian service. Under the terms of this agreement, the Prussian corps in Courland left Macdonald's X Corps and became neutral. This *fait accompli* forced Friedrich Wilhelm to take the decision to break openly with Napoleon and to embark on a desperate gamble to defeat the hitherto undisputed military and political master of most of mainland Europe. In any case, by the end of December the scale of the destruction suffered by the Grande Armée in Russia was becoming known. There was little danger of immediate reprisals being taken against Prussia.

The Clash at Haynau

All of Europe was now plunged into a frantic arms race to replace the men, horses and materiel that had been lost in 1812. Austria declared itself to be neutral, but rearmed as fast as it could.

The spring campaign in Saxony began to get under way in April, with a series of mainly minor clashes as the opponents closed up to one another. On 2 May Napoleon won the Battle of Lützen (Gross-Görschen) about 12 miles south-west of Leipzig but, thanks to their superior cavalry force, the allies withdrew to the east unmolested, abandoning Leipzig and Dresden. Napoleon followed up as fast as he could. Both sides lost about 20,000 men.

On 20/21 May he again defeated the allies at the Battle of Bautzen on the River Spree, 30 miles north-east of Dresden. The allies continued to withdraw eastwards. Next day they were again beaten in a clash at Reichenbach, seven miles west of Görlitz, but held the French back until they had crossed the River Neisse and burned the bridge.

There was discord in the allied camp; they blamed each other for the defeats. Eventually, Wittgenstein would grow tired of being the scapegoat for Alexander's temper and would resign as allied commander-in-chief. Barclay de Tolly proposed that the Russian and Prussian forces should withdraw into Poland. Blücher vehemently opposed this plan, as it would abandon Prussia to the French. Leaving the Army of the North to protect Berlin, the allies decided to take up a position to Napoleon's right flank around Schweidnitz, west of the upper Oder.

By 24 May the main body of the Army of Silesia had crossed the River Bober and passed through Bunzlau. Next day this force reached Haynau. The French advanced guard, Maison's 16th Division of Lauriston's V Corps and Chastel's 3rd Light Cavalry Division of I Cavalry Corps, followed slowly. Maison had only four regiments of new conscripts, which had all been raised on 12 January 1813. Of these, the 151st, 153rd and 154th had taken heavy losses at Wiessig on 19 May and were weak.

Wittgenstein's Russians were now between Löwenberg and Goldberg to the south. Blücher's rearguard of three battalions of infantry and three light cavalry regiments (only one of which was present on the day of the action) was commanded by Colonel Mutius. At this point Blücher received orders to hold up the

The Clash at Haynau

enemy if they should try to advance further than Haynau. The army was to try to make a stand again behind the River Katzbach. Blücher relished the task. Usually he left strategic planning details to his chief of staff, Gneisenau; this time he took over himself. He rode around the area for hours, questioned the inhabitants, sent out scouts and then returned to his staff to discuss his plan with them.

Blücher outlined the situation:

> 'The aim is to lure the enemy out of Haynau, into the low plain between the villages of Überschaar and Pohlsdorf; then to attack him with concealed cavalry and artillery and cut him off from his retreat back into Haynau.
>
> The 23 squadrons of Colonel von Dollfs's Reserve Cavalry and the horse artillery batteries will conceal themselves between Baudmannsdorf and Überschaar. Colonel Mutius's rearguard will come through Steinsdorf and Haynau and on towards Pohlsdorf, the same way that Colonel von Pirch's infantry took.
>
> Pohlsdorf is the point which must be held if all goes wrong. If the enemy advances, General Zieten will advance to within 500 paces, attack his column, and when it is in confusion, will light a fire on the windmill hill at Baudmannsdorf, as a signal for the cavalry to march off to the left, pass round the enemy, cut them off, cut them down, etc.'

The plan worked perfectly, unwittingly aided by the French. There seems to have been no co-ordination of the advance on their part. It was late afternoon when they reached Haynau and Chastel considered the day's work done and went into bivouac with his cavalry. Either in ignorance of this, or out of sheer foolhardiness, Maison pushed on alone with only his infantry. He was possibly stung into foolhardy action by Ney, who was initially with him. As they advanced through Michelsdorf, Maison questioned if it was wise for him to push on without having scouted ahead. Ney laughed and retorted that the Prussians would not stand. Thereupon, he returned to Haynau. Maison continued his advance.

The unsuspecting prey defiled through Michelsdorf and along the Liegnitz road, following Mutius's rearguard. From the top of

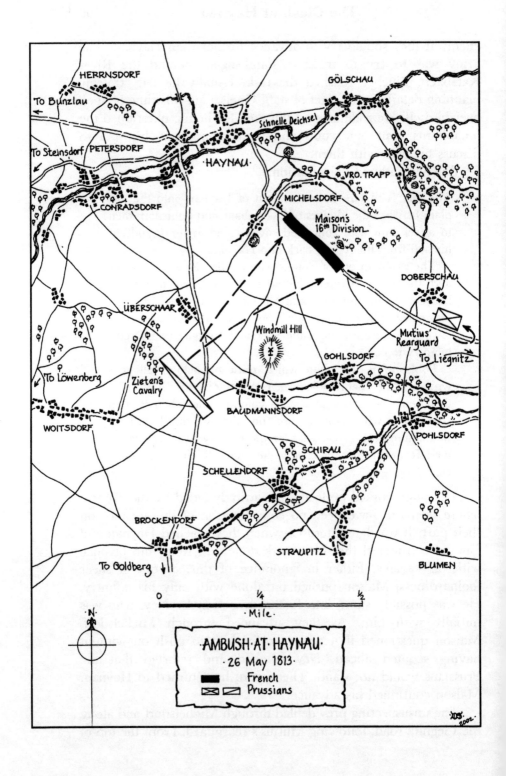

The Clash at Haynau

the windmill hill north of Baudmannsdorf, Blücher and his staff waited in concealment as the enemy snake worked its way into the trap. With no cavalry scouts, General Maison's vision was strictly limited. When the head of his column reached Doberschau, things suddenly started to happen. Mutius's rearguard formed across his front and a cloud of cavalry and horse artillery poured over the low ridge to his right about 1,000 yards away. As French attention was fixed on Mutius, it was some moments before the threat from the right was noticed.

The windmill suddenly burst into flames. Enemy cavalry! Maison had none! Withdrawal into Haynau was impossible so Maison ordered his regiments to form square, ready to receive cavalry; it was all he could do. The Prussians trotted quickly over the open meadow, taking care not to obscure the field of fire for the guns with them. The artillery unlimbered and came into action at close range, pouring canister into the French squares. For a time they held firm, then the Prussian cavalry hit them; first the Silesians, then the East Prussian Kürassiers. One square collapsed, then the other; the cavalry had a field day. Now Mutius's infantry were coming up. Despite all his efforts, Maison's command crumbled before his eyes. He was swept back towards Haynau in the tide of panic-stricken infantry, his division now a wreck.

Hearing the gunfire, Chastel at once had his men saddle up and rush out to the clash. Only part of his command actually came into action and only the 25th Chasseurs show any officer casualties for this day – one wounded. The chase in fact went on into Haynau. There was panic in the French camp. The surprise was complete, the victory absolute.

French losses were 1,400 killed, wounded and captured; 15 guns were taken, but only 11 could be taken away. The French admitted losses of 1,350 men and five guns. The 151st Line lost three officers killed, one mortally wounded and 14 wounded. The 153rd Line lost two mortally wounded and five wounded. The 152nd and 154th had no officer losses. The Prussians lost 226 men and 205 horses killed, wounded and missing. The Silesian Kürassiers had cut down the 151st and taken four guns for the loss of four men killed, 15 wounded and 58 horses wounded. The East Prussian Kürassiers had ridden down two squares of the

153rd Line for the loss of one man and six horses killed, four men and two horses wounded. The Brandenburg Kürassiers lost three men and seven horses wounded. Colonel von Dolffs was among those killed. The infantry were not heavily engaged.

Blücher had won the time that the allies needed. For this day's work he received the Order of the Iron Cross, 2nd Class, and the Russian Order of St. George, 2nd Class. General-Major von Zieten won the Iron Cross, 1st Class, for his part in this action. This was the last action of note before the Armistice of Poischwitz came into force from 4 June to 22 August.

Odeleben, a Saxon colonel and ADC to the king, accompanied Napoleon through this campaign. This is what he had to say on the ambush at Haynau. The casualty figures that Odeleben gives are probably those that the French fed to the Emperor.

> 'On 25 May, the Emperor left Görlitz at midday and moved his headquarters to Bunzlau, which his troops had occupied that same morning. He stayed there until 26 May.
>
> Marshal Ney was pushed forward to Hainau and lost some battalions that evening.
>
> Enemy cannon fire had completely demolished a pair of squares composed of new conscripts, and one of them was cut down by the Prussian cavalry.
>
> Very early next morning, Napoleon hurried to Marshal Ney in Hainau. He rode at a gallop, such as he had not done since the day of Naumburg, in order to view the battlefield.
>
> If he was depressed about suffering a defeat, his generals usually sought to lift his spirits by emphasising the enemy's losses.
>
> This was the case here. They had removed all the French dead, and only the Prussian corpses still lay about. One of them was identified to him as Lieutenant-Colonel von Buchholz, who had led the charge.
>
> Apart from six guns, the French lost about 800 men dead, wounded and captured.
>
> It was again proved that – up to this day, which was the last occasion on which anything of note took place before the Armistice – the allies always lost fewer men and guns than the French Army. This was apart from the allies' advantages

The Clash at Haynau

of outflanking moves with their cavalry and the capture of despatches and couriers.

At Rippach and Lützen, Königswartha, Bautzen and Reichenbach – everywhere the French losses were greater, and we could not compensate for this with captured cannon or other trophies, nor with enemy generals that we had killed.

On the contrary, Marshals Bessières and Duroc had been killed, as had Generals Delzon and Grüner at Lützen, and Bruyères and Kirchner at Reichenbach.

The numbers of the dead were increasing, and I was convinced that the deaths of many senior – and other – officers were being concealed from us.'

Sources

Chandler, David, *The Campaigns of Napoleon*, London, 1966

Esposito, V.J., and Elting, J.R., *An Atlas of the Napoleonic Wars*, London, 1999

Martinien, A., *Tableaux pars Corps et pars Batailles des Officiers Tués et Blessés Pendant les Guerres de l'Empire 1805–1815*, Paris, 1890

Müller-Boehm, Hermann, *Die Deutschen Befreiungskriege*, Berlin, c.1913

Odeleben, Otto, Colonel, Freiherr von, *Napoleon's Campaign in Saxony in the Year 1813*, Dresden, 1816

Priesdorf, Kurt von, *Soldatisches Führertum*, Vol II, Hamburg, 1937

Six, Georges, *Dictionnaire Biographique des Generaux & Amiraux Français de la Revolution et de l'Empire (1792–1814)*, Paris, 1934

Voigt, Günther, *Deutschlands Heere bis 1918*, ten volumes, Osnabrück, 1983

Chapter 10

The Clash at Liebertwolkwitz

13–14 October 1813

The spring campaign of 1813 showed Napoleon at his best again, combating superior – but scattered – enemy forces with his wits and his men's legs as in the past. His command and staff structure down to brigade level had survived the Russian campaign pretty well intact and sufficient field officers remained capable of active service to allow the regimental level of command to be reconstructed by promoting from the surviving company level officers. Conscription filled the lower ranks with sufficient cannon fodder for his immediate purposes, although resistance to this 'tax of blood' was growing in France as well as in the allied states. Weapons, equipment and vehicles could be manufactured relatively quickly; replacing cavalry and train horses was a very different matter. In Russia Napoleon and his allies had lost over 95 per cent of their horses, and nature would have to take its course before large numbers of new foals could be sired, born and grow to working age. This was a problem that confronted all the belligerents of course, not just the French.

Despite all these difficulties, the Emperor was able to take the field in strength again in mid-April and to advance on Leipzig to maintain his iron grip on Germany. The first major battle of the 1813 campaign took place at Lützen (Gross-Görschen) on 2 May; the next was at Bautzen on 20/21 May. Both these actions were victories for Napoleon, largely due to the inability of the allies to make the best use of their superior cavalry forces. They were, however, not crushing defeats of the type of Ulm, Austerlitz or Jena-Auerstädt. His enemies had studied busily in the Emperor's hard school of war since then. They had learned, and were applying, some valuable techniques. Even so, by the conclusion of these battles, both sides were already shaken by the exertions that they had been forced to make. An armistice was agreed on 4 June. It lasted for seven weeks, during which time frantic military refurbishment vied with feverish diplomatic activity as alliances were massaged into place.

On 12 August Austria entered the lists on the allied side, and hostilities resumed. There followed a period of urgent military manoeuvring across the devastated terrain of Saxony as each side sought to deliver the knockout blow to the other. Inter-allied politics and rivalry played their part in limiting the allies' effectiveness against the greatest general of their time. During the armistice, however, they had agreed an operational strategy. They had three armies in the field: the main Army of Bohemia under Schwarzenberg, the Army of Silesia under Blücher, and the Army of the North under Bernadotte, the Crown Prince of Sweden. These three armies were distributed on the edges of a ring around the Grande Armée in the Dresden–Leipzig area. Their so-called 'Trachenberg Plan' was very simple: whichever allied army was attacked by Napoleon would retreat; the other two would close in to operate on his flanks. The aim was to run his men into the ground with pointless marching and counter-marching. It worked.

For his part, Napoleon never realised what was going on; all he knew was that his army was being worn out, chasing evasive opponents around Saxony. In reprisal, he launched repeated jabs at Berlin; they were all beaten off. On 22 August the allies won the clash at Pirna (Goldberg); next day, they won again at Gross-Beeren. Then Blücher gave the Allies their best victory of this phase of the war in the battle of the Katzbach on 26 August, where he soundly defeated Macdonald. Of course, the Allied commanders blundered every so often, badly so at Dresden on 26 and 27 August, where the Army of Bohemia was severely mauled, but they were able to keep their forces in operational condition.

The advantages which Napoleon achieved at Dresden were squandered over the next few days by his generals at Plagwitz (Bober River) on 29 August, Kulm on 29/30 August and Dennewitz (Jüterbog) on 6 September. If one analyses Napoleon's movements around the theatre of war in this period, they resemble nothing so much as the bumblings of a great French bee, caught in the toils of the web of the Trachenberg plan. He was also chronically short of reliable intelligence as to the actual locations and intentions of each of the three allied armies. This was a direct result of his inferior cavalry and their exhaustion.

While the French soldiers became weaker, their equipment wore out and their few horses needed shoeing more frequently, each allied army was able to enjoy the odd 'rest period' when the imperial heat was concentrated on one of its allies. Co-ordinating the movements of the three separated armies in 1813 was not easy, however. They were operating on exterior lines and obviously had no radio communications to help bridge the gaps between them. Added to this was the fact that Bernadotte had immense respect for Napoleon's battlefield talents and kept well out of his path. He was also determined to protect his precious new countrymen at all costs, and equally determined not to inflict too much harm on his old comrades in arms. All in all, the Army of the North was not much help to the allied cause at all in 1813, and it took all of Blücher's efforts – and more than all his patience – to win even minimal help from it at the vitally important Battle of the Nations at Leipzig.

More clashes followed after Dennewitz, but the most significant occurred on 3 October, when von Yorck's I Prussian Corps of the Army of Silesia forced the crossing of the Elbe, heralding Blücher's offensive that was to culminate in the devastating allied victory at Leipzig. On 8 October, after much haggling with Austria, Bavaria abandoned Napoleon and joined the allies by the Treaty of Ried. The stage was now almost set for the great, decisive battle.

Over the last few weeks, Napoleon had been in uncharacteristically indecisive mood. He had lost the initiative due to the workings of the Trachenberg Plan and he knew it. He dithered as to whether he should hold on to Dresden or abandon

The Clash at Liebertwolkwitz 167

it, and on 6 October explained to Gouvion St. Cyr his excellent reasoning for the city's abandonment. A couple of days later, he changed his mind and put two corps in the place as garrison. As he himself had just said: 'I shall need those men in the field.' He did. As he himself had just said: 'They will surely be lost.' They were.

On the evening of 13 October Murat, who was now commanding Napoleon's southern flank forces, aimed to withdraw northwards over the Parthe and to stand between Leipzig and Taucha. In doing this he would have given up the best terrain for the forthcoming battle and exposed Leipzig itself to attack. As soon as he heard this, Napoleon sent an aide, Major Baron Gourgaud, to stop him. Gourgaud found Murat already withdrawing and delivered the Emperor's message. Murat then turned back and took post on the slight but dominating heights between Markkleeberg and Liebertwolkwitz. The fields of fire from here were superb; the rear and lateral communications very good. Murat's right wing was on the Pleisse/Elster swamps. His dispositions were as follows:

Right wing: Markkleeberg–Dölitz–Lössnig–Connewitz – Kaminiecki's Division of Poniatowski's VIII Corps – 5,400 infantry, 600 cavalry, 30 guns

Centre: Between Markkleeberg and Wachau – Victor's II Corps – 15,000 infantry, 38 guns

Left: From Wachau to Liebertwolkwitz – Lauriston's V Corps – 12,000 infantry, 700 cavalry, 53 guns

At Holzhausen, as reserve, was a division of the Young Guard. Further back, at Thonberg, was Augereau's IX Corps (9,500 men), which had just arrived from Spain. Also present was Milhaud's 6th Heavy Cavalry Division, followed by l'Héritier's 5th Heavy Cavalry Division and Berckheim's 1st Light Cavalry Division (of I Cavalry Corps), commanded by General Pajol. This 4,000-strong cavalry force was set behind V Corps between the Galgenberg and the Monarchenhügel (Monarch's Hill). IV Cavalry Corps (commanded by General Sokolnicki, as Kellermann was sick) was behind VIII Corps at Markkleeberg. These French and Polish regiments fought on 14, 16 and 18 October and suffered the heaviest losses of the Grande Armée.

These units were opposed by Wittgenstein's Russians and

Kleist's Prussians. On the 14th, Wittgenstein had Prince Gorshakov's I Corps, Prince Eugen of Württemberg's II Corps and General Pavel Petrovich Pahlen's cavalry (Illovaiski XII's Cossacks, the Hussar brigades of General Rüdiger and Colonel Schufanov and General Lisanevich's Ulan brigade). Kleist's II (Prussian) Corps consisted of four infantry brigades and General Röder's Reserve Cavalry Brigade. The Allies had about 6,000 cavalry, 20,000 infantry and 48 guns on the field by the end of the day. They were opposed by about 7,000 cavalry, 32,400 infantry, and 121 guns.

What now developed was the largest cavalry clash of the 1813 campaign. First contact was between Illovaiski's Cossacks and the French cavalry of l'Héritier and Subervie near Liebertwolkwitz. The newly arrived dragoons from Spain were apparently distinctive in their brown greatcoats which had been made up in Spain. Between 9 and 10 o'clock Pajol withdrew into the main French position between Wachau and Liebertwolkwitz and the forward French tirailleurs evacuated Gross-Pössna and fell back to Liebertwolkwitz. Pahlen advanced on Magdeborn; Klenau's Austrians arrived in Gross-Pössna at 9 o'clock and Feldmarschall-Leutnant Mohr's 1st Division took the village.

General Pahlen had expected to snatch an early victory but was overthrown; the Soum and Grodno Hussars were driven off in confusion. Pahlen sent to Kleist for help, but before the Prussian cavalry arrived, Prince Eugen's Russian infantry came up by Güldengossa.

On the left wing at Auenhain was Illovaiski with three Cossack *pulks* and the Russian Guard Hussar Regiment. On the right wing, on the meadows north-east of Güldengossa, were the Lubny and Olviopol Hussar Regiments. In the second line were the Chuguyev Ulans. In the Allied centre on the heights north of Güldengossa were the Soum Hussar Regiment and a Russian horse artillery battery. This battery fired on the French cavalry; two of the enemy regiments charged the battery, which only just managed to limber up and escape in time. The Soum Hussars counter-attacked but were outnumbered and had to flee into Güldengossa.

At this point, the Prussian Neumark Dragoons arrived. Pahlen greeted them with 'Just get at them, we'll overthrow them

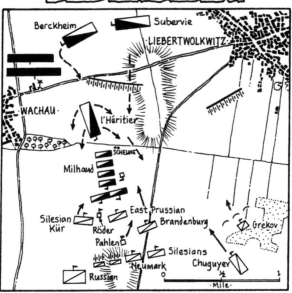

THE ADVANCE OF THE PRUSSIAN RESERVE CAVALRY

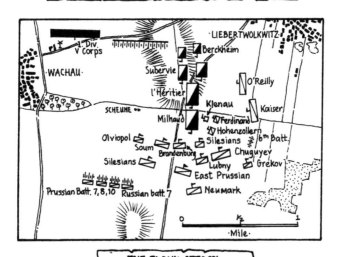

THE FLANK ATTACK BY THE AUSTRIAN CAVALRY

at once!' The Prussians answered: 'None of us is scared!' and charged forward together with the Russians. The fight was hectic, bitter.

Murat now threw in Milhaud's dragoons with l'Héritier's and Subervie's men in second and third lines. Some 5,000 cavalry advanced in an apparently invincible mass. Milhaud's division was deployed as follows: at the front of the column were the 22nd and 25th Dragoons, beside one another and in line abreast; behind them rode the 20th, 19th and 18th, each in line abreast. At this critical point, General Röder came up with the Brandenburg, East Prussian and Silesian Kürassiers. The heavy French phalanx was hit in the right flank. Due to their close formation, most of the French troopers could not use their weapons and the whole mass was pushed back to Wachau. There, however, the three Prussian Kürassier regiments were confronted by French infantry and surrounded by the enemy's hussars and chasseurs-à-cheval.

There was no plan. The regiments were simply thrown into the mêlée as they came up. It was chaos, friend and foe mixed together. The fight was so intense that often the exhausted men and horses rested close to one another before starting to fight again. Gradually, after a hard struggle, the Prussians were beaten back but were taken up and rallied by the Neumark Dragoons.

Murat was in the thick of it. Distinguished by high ostrich plumes in his hat, and a gold-embroidered velvet costume, he came very close to being captured. Sources differ as to which officer it was that nearly took him. Some say it was Leutnant von der Lippe of the Neumark Dragoons; Zelle states that it was Major von Bredow of the Brandenburg Kürassiers. Whoever it was, was close behind Murat at one pont, shouting 'Halt, König, halt!' when Murat's master of horse shot him down from behind.

On the evening of 13 October, Klenau had been at Borna, Lausigk, and Pomssen. On the 14th, Klenau's main body got to Threna. Here he received Wittgenstein's order to take Liebertwolkwitz and to push on to take Murat's cavalry in the left flank. Klenau hurried on and, at 11.30 on the 14th, his Infantry Regiment Erzherzog Karl Nr 3 assaulted Liebertwolkwitz. Klenau had sent 15 squadrons of cavalry under

The Clash at Liebertwolkwitz

General Desfours from Threna on to Güldengossa to support his allies. They arrived in the nick of time at 12:30.

At 2 o'clock in the afternoon Murat concentrated his cavalry for another assault. Milhaud's Spanish veterans were again at the head of the column, l'Héritier's dragoons behind them, then Subervie and then Berckheim. The attack went due south. Like a great, shining snake, the massive column of horsemen burst out of the smoke and bore down on the Allies. As the Russian Adjutant Molostov recalled:

> 'All shrank back from this glistening vision which embodied for us the magic that surrounded Napoleon's brows. The mass of riders, with the sun glancing from their weapons and helmets, formed one huge, endless column which crushed all before it and hit the Prussians particularly hard.'

But the fire of the four Russian and Prussian horse artillery batteries on the heights north of Güldengossa (the 7th, 8th, 10th Prussian and 7th Russian) hit the front and right flank of the column and ripped its head apart, stopping it in its tracks. Seizing the opportunity, the Russian hussars, with the Prussian Ulans and the Brandenburg Kürassiers, charged into the disorganised front ranks. At the same time, Desfours led the fresh Austrian cavalry (Kaiser Kürassiers, O'Reilly and Hohenzollern Chevauxlégers and, according to one account, the Ferdinand Hussars) into the mêlée. They advanced from the bend in the Güldengossa-Liebertwolkwitz road and ploughed into the left flank of the closely-packed enemy column. This completed the confusion of Milhaud's division and threw it back onto l'Héritier's men. In a few minutes Murat's great column burst apart in all directions in panic and fled the field. They could only be rallied at Probstheida, some two miles to the north. It was now about 4 o'clock.

At the same time that the Austrian cavalry launched their charge against Murat's column, the Austrian infantry attacked Liebertwolkwitz. The Infantry Regiment Erzherzog Karl under Colonel von Salis advanced from Niederholz via Gross-Pössna against the eastern side of the village. It was soon involved in heavy fighting with Maison's 16th Division, which defended the place stubbornly from house to house. By about 2 o'clock the

Austrians had taken the stone-walled *Gottesacker* (churchyard) after a bloody combat. Maison, aware of the vital nature of the position, counter-attacked with the 153rd Line and drove the Austrians out. The Austrians regrouped and assaulted again with General Splenyi's brigade (the Württemberg and Lindenau Regiments) and a battalion of the Wallachisch-Illyrisch Grenz Infantry Regiment and retook the cemetery. Maison threw in reinforcements (the 152nd and 154th) and wrested the disputed point back. It was now about 5 o'clock. The fighting was so violent that no quarter was given. The Austrians were pushed back against the wall around the cemetery gate, which opened inwards. The crush was so great that they could not defend themselves. Next day witnesses found rows of dead Austrians, leaning against one another, skewered to the wall by bayonets.

The Austrian assault on Liebertwolkwitz from the south was supported by General Paumgarten's brigade from the area of the Kolmberg to the east. This thrust got as far as Pössgraben, where it was halted by the fire of a French battery deployed to the north of Liebertwolkwitz. There was no further action on this wing.

The situation was very different on the western flank, up against the River Pleisse. Here General Illovaiski's Cossacks and von Rüdiger's Grodno Hussars advanced north from Auenhain on the gap between Markkleeberg on the eastern bank of the river and Wachau about 1,200 yards to the east. Poniatowski threw his entire cavalry force against them and the heavily outnumbered Russians fell back south to Cröbern.

Now it was the turn of the Prussians. Colonel von Mutius, with four squadrons from the 1st, 7th and 8th Silesian Landwehr Cavalry Regiments, charged the west wing of the Polish cavalry line and, supported by some of Russian General Helfreich's Jägers, drove them back into their main position. However, the fire of a powerful French battery, deployed on the heights east of Wachau, stopped any further Allied advance in this sector.

It was now about 6 o'clock, and Prinz Schwarzenberg, who had arrived on the scene, ordered the action to be broken off. There was little further offensive action by the French and Poles either, as General Duka's 3rd Cuirassier Division had now arrived at Cröbern to reinforce Illovaiski and Rüdiger. A stalemate ensued.

The Clash at Liebertwolkwitz

The Allies lost the Russian Lieutenant-General Docturov and Major-General Nikitin killed, 2,500 officers and men killed and wounded, and 500 captured. The Silesian Kürassiers lost 14 officers and 164 men, the Brandenburgers six officers and 45 men, the East Prussians ten officers and 52 men, the Silesian Ulans ten men, the Neumark Dragoons about 50 men. The Austrian cavalry lost about 70 men, all ranks. Klenau's infantry lost 867 killed and wounded and 134 captured in the bitter fighting in Liebertwolkwitz itself. According to Napoleon, the French lost only 400 to 500 men in all, whilst the Allies lost 1,200 prisoners and two guns, without counting their dead and wounded. In fact losses suffered by the French during the action comprised about 1,500 killed and wounded, and 1,000 captured. General Bertrand was among the wounded.

Thus ended the action of 14 October. Murat had achieved his given task of holding the dominant heights between Markkleeberg and Liebertwolkwitz, but the cost had been heavy and most of these fell upon the precious cavalry of which Napoleon had so little. The fault for this was Murat's alone. The use of a closely-packed, heavy column of cavalry in the face of the close-range fire of four batteries of hostile artillery – which were allowed to operate without any attempted disruption from the French side – invited a costly defeat. The whole sorry episode demonstrated Murat's complete failure to co-ordinate the actions of the assets available to him. In some ways, it was a foretaste of the needless sacrifice of the French cavalry at Waterloo in 1815.

The Russian General von Toll judged Murat's influence on this clash, and his abilities as a cavalry commander, as follows:

> 'This singular combat lasted so long thanks only to Murat's thoughtless pugnacity. We may mention here that the reputation of this theatrical monarch as a splendid cavalry commander, "the Seydlitz of Napoleon's army", was undeserved and survived only because nobody could tell the truth about Napoleon's brother-in-law.
>
> Murat was completely incapable of leading large formations of cavalry. Generals commanding other army corps sought to hide their cavalry brigades from his view when he was in their vicinity, for if he was aware of them, it was quite likely that he would commandeer them and ruin

them in some senseless brawl. If he was in command of large cavalry masses, he very easily lost control over them as he could only manage what went on in his immediate area. Men like Latour-Maubourg, Nansouty – and above all the very capable Montbrun – knew how to look after themselves, especially when he was not in their vicinity to hinder their efforts.

This time he thought it fun to take command of the veteran dragoon regiments which had returned from Spain with Augereau and these worthy warriors upheld their reputations for sustained gallantry completely. This combat went much against the French, however, mainly – as German officers related – due to the fact that their own horses were in much better condition than those of the French. This allowed them to mount charges late in the action with much the same power and energy as at the beginning, whereas the French dragoon mounts were blown and weak. Thus these capable, veteran dragoons, who had so much irreplaceable combat value, lost about a third of their strength including 500 captured.

Finally, Murat pulled them back under cover of the French batteries where Pahlen – who would willingly have broken off the fight earlier – could not follow them.'

Edouard von Löwenstern, Pahlen's ADC, related how he saw the fight:

'On 13th October, Count Pahlen advanced with all his cavalry, reinforced by several Prussian regiments, to Cröbern. On the 14th October the Soum and Lubny Hussars were the advanced guard; they were followed by the East Prussian and Neumark Kürassiers and the Silesian Ulans. Right up front was the brave Markov and his guns.

Scarcely had he unlimbered and begun to fire in earnest, than the enemy cavalry moved off towards the battery at a trot.

Not far from Güldengossa and the village of Liebertwolkwitz, we made contact with Marshal Augereau, who had led 10,000 dragoons up from Spain, fine, strong men on excellent Andalusian horses. With the sun glinting from their helmets and swords, this terrible phalanx approached us. Markov received them with well-aimed fire.

The Clash at Liebertwolkwitz

The French deployed 50 paces from us and charged at us. Lances, sabres and swords clashed against one another. Many were thrown from their mounts in the first shock and trampled underfoot. At one moment we were charging forward with a great "Hurrah!", the next we were riding for our lives and the "Hurrahs" had become cries of fear. We overthrew them and were overthrown in our turn. To right and left, to front and rear, all we could see was our men and the enemy hacking and stabbing at one another, with neither side getting the upper hand. The Prussians remembered that this was the anniversary of Jena and fought like lions to wipe out that disgrace.

Count Pahlen was always at the head of his regiments, between the cavalry lines of the enemy and his own side. He led the charges himself and directed the reserves. He was always in the thick of the fight, where it was most critical, with only a light riding whip in his hand, giving orders calmly.

If I had been an enthusiastic admirer of his in the past, I now had to honour and admire him even more. This was not the blood-crazed, berserk bravery of a Figner, or the half-mad courage of a Seslawin [Figner and Seslawin: Russian war heroes]; this was the cool-bloodedness and noble bravery of a great commander. This day alone must have made his reputation as an extraordinary general.

Both sides fought with great bitterness. The Prussian Kürassiers were already in possession of a French battery and were preparing to tow away some of the guns, when fresh troops came up to stop them.

This was how I imagined ancient battles to have been – man against man – the stronger could be sure of his victory.

I don't know whether it is courage or the need to find a quick death and so to escape the continual fear of death, which drives one on. Foaming at the mouth, one charges rashly into the enemy formation. Some hussars and enemy horsemen had died clenched together in their final efforts.

A French officer on a grey, all on his own, rode up to the Count with his sabre raised; 'F..., surrender!' he shouted. A Soum Hussar lunged at him with his lance. The Frenchman turned his horse and tried to escape but was cut down before he reached his own lines.

As Count Pahlen had received orders not to let himself be drawn into a serious fight, towards evening the two sides pulled apart. Now the artillery started up again and continued to thunder until darkness fell.

The casualties of the day lay together: French, Russians and Prussians, covered with gaping wounds from the edged weapons. Our cavalry reformed and went into bivouac by Cröbern.

The East Prussian Kürassiers had covered themselves with glory.

The King of Naples had commanded the enemy cavalry and had made much use of General Pajol's dragoons, which had only come back from Spain in the summer under Marshal Augereau's command.

The French had learned to respect our cavalry again this day, and they did not dare to engage us in such large formations again on the following days of the battle.

This bloody mêlée cost the enemy well over 1,000 prisoners, including many officers, and about as many again dead. Our hussars had taken 518 men prisoner. We lost about as many dead as the enemy but lost no prisoners.

The Soum Hussars lost 11 officers, including Schischkin, my old squadron commander, who lost a leg. Count Ivan Pahlen received a pistol shot in the leg.

Next day, everything was quiet; none of us wanted to start the witches' dance again, except the Cossacks of Grekov and Illovaiski, who were on outpost duty.'

Ernst Maximilian Hermann von Gaffron (ennobled in 1840 as the Freiherr von Gaffron-Kunern), a Portepee-Fähnrich (officer aspirant) in the Silesian Kürassiers, has left us his recollections of this dramatic day:

> 'The morning of 14 October was cold and foggy but by midday the sun had broken through and it became a little warmer. I went looking for our regimental commander, Major von Folgersberg, my faithful mentor and a friend of my father's, and found him lying on one of the regimental baggage carts, covered with a blanket and shaking with a violent fever. He had been feeling unwell for some days. He spoke to me in a weak voice. When I saw his condition, I told him that it was impossible for him to stay with the

The Clash at Liebertwolkwitz

regiment. He answered me with a weak grin: "You can't believe that I am going to leave the regiment now? We might be in a battle today or tomorrow. I am not going to give up the honour of commanding the regiment on such a day, even if it kills me!"

He would not be moved from his decision; I felt that I would probably do the same. He shook my hand firmly and spoke of my father, then we parted. I met him once more, in the middle of the battle, but these were the last words I heard him speak, apart from: "March, march, charge!"

It must have been between 10 and 11 o'clock when an ADC galloped up and the order rang out: "To your horses and mount!" We moved off but soon came the command: "Trot!" We trotted for about an hour, at first in column, then in troops, across country. We came to a bivouac site where the straw, huts and campfire embers told us that the French must have left the place only that morning. These traces, and our rapid trot, told us that we must be near the enemy.

Röder's cavalry [the three Kürassier regiments, the Silesian Ulans and the Neumark Dragoons – 18 squadrons in all, as two squadrons of Ulans were detached – two horse artillery batteries, and some weak Volunteer Jäger detachments] had been placed under command of the Russian General Count Pahlen [II]. He also commanded the Russian Grodno and Soum Hussars [12 squadrons] and a Russian horse artillery battery and thus had thirty squadrons and three batteries.

The Russian General Count Wittgenstein, who commanded our Army Detachment [the corps of Wittgenstein, Kleist and Klenau], had ordered a reconnaissance in strength on Liebertwolkwitz to clarify enemy dispositions on the far side of the Pleisse. Pahlen's cavalry was the advanced guard, of which the point unit was the Grodno Hussars and ourselves.

In order to gain the gentle heights of Wachau and Liebertwolkwitz, we had to pass through the defile of the pretty village of Cröbern, which lay in a shallow valley. This we did at a fast trot, with our carbines at the ready. As we passed through, I noticed the fine, friendly-looking manor house, which lay to one side and was a picture of peace and cosiness.

Once through the village, we at once went into a gallop, first in troops, then on a regimental frontage. We halted briefly, then came the call "Trot" and we moved off smartly again.

A dark mass stood against us, to their right front shone the white tower of Liebertwolkwitz. With my short-sightedness, I could not make out the details of the dark mass, but soon I saw the flashing of helmets and swords as they closed with us.

The signal "Gallop" sounded, and now I could see the enemy squadron clearly. The regiment was going at such a pace that even before the command "March, march!" ["At the double"] came, a thunderous cheer broke from our ranks. Folgersberg, who was riding on the left wing of our squadron, raised his sword, his "March, march!" was echoed in our hurrah and the trumpet signal. The French dragoons, whose moustachioed faces I could now clearly see, stopped, suddenly turned, and withdrew in good formation.

We soon caught up with them as they were only about 20 paces before us. The sight of the enemy cavalry had brought the regiment to boiling point. This was the anniversary of Jena; that blot had to be wiped out! This was our first cavalry combat of the campaign, something we had been waiting for for a long time.

An ADC, who witnessed our charge, told me later that it was a splendid sight. First the fresh tempo of the trot – in a parade-ground dressing – then the full gallop, then the impact as we hit the enemy and swept them away as would a storm.

The enemy dragoons were from Milhaud's first rank. As far as I remember, the Grodno Hussars were first through the Cröbern defile, then our regiment. Since we went straight into the attack I only recall seeing men of these two regiments. The Grodno Hussars were, as usual, brilliant in this combat. The other regiments followed us as quickly as they could clear the defile.

At the point at which we cheered in that charge, and we could see the whites of the Frenchmen's eyes, there was no stopping us. Every rider spurred his horse and sought to split a French skull.

Our dear commander Folgersberg rode in front of the interval between the 1st and 2nd Squadrons; I was on the

The Clash at Liebertwolkwitz

right wing of the 4th Troop, 1st Squadron. As we cheered, I whooped with joy, gave my chestnut my spurs and was one of the first to reach the French, who turned about at this instant.

There was no order in the ranks any more, everyone rode like the wind. I passed Folgersberg at about six paces distance as he rode his fine, great iron grey. He nodded and waved to me in a friendly, but warning manner. That was our farewell. As I broke forward out of the ranks of my troop, the faithful Czekyra, who rode behind me shouted: "Gently, gently, Herr Portepee-Fähnrich!" But his warning fell on deaf ears.

The French dragoons retired before us, initially at a trot, then at a gallop, in good formation. We often heard the officers calling: "Serrez les rangs!" ["Close ranks"] and in fact, they rode so closely together that at first we could not break their ranks, but stayed close on their heels and hacked at them as best we could. The horse-tail manes of their helmets, their broad, stout leather bandoliers and the rolled greatcoats, which they wore over their shoulders, protected them so well that they were pretty impervious to cuts, and our Silesians were not trained to thrust nor were our broad-bladed swords long enough to reach them.

Just hacking at the heads and backs of the enemy did not do much good. It was not until some of our cleverer men knocked some of them out of their saddles with well-aimed stabs or pistol shots, that we were able to make a gap in their phalanx. Finally, we managed to break into their ranks and they burst apart. As our regiment was now out of order, the combat developed into a series of small, fighting groups in which the French lost many men.

At this point in the battle, I bumped into my young comrade, Senft von Pilsach, who had been made Portepec-Fähnrich with me a few days before. "Well, Senft, today is a fine day!" I called out to him; he smiled and nodded to me. A few moments later, he sank to the ground, his skull split by an enemy sword.

We chased and fought with the enemy for a long while. We threw his first line back onto his second, and his second onto his third. Suddenly, the enemy's ranks broke apart in front of us. A battery of horse artillery had become jammed in the mess of the dragoons and was soon surrounded by our men. We cut down the gunners and the drivers, or forced the latter to turn the guns around. We had turned several of the guns around, and I was about to take them back with some of my comrades, drunk with the joy of victory and oblivious to what else was going on around me, when suddenly I heard: "En avant, dragons, à-bas ces foutous Prussiens!" coming from all sides!

I then realised that I and about 20–30 Kürassiers in the captured battery were surrounded by hostile dragoons. A colonel of the enemy dragoons, a tall, handsome man on a large, English thoroughbred light chestnut decked out in the finest equipment, called his dragoons together and charged us.

The captured guns were abandoned. Our small group of Kürassiers would have to cut our way out. We had to try to go it alone as there was no chance of forming a tight group. Some Kürassiers were behind me. I heard one cry "Jesus and Mary!" as one of them was cut down.

Five or six dragoons were after me. I thought I was lost, but swore to try everything to escape. To my horror, my chestnut began to lose speed, be it due to fatigue or for some other reason, despite the fact that I was urging it on with my spurs and the flat of my sword. As I was wearing an officer-pattern greatcoat and cartouche, the dragoons thought that I was an officer and chased after me all the harder.

The ground that we rode over was unploughed, very wet and very hard going for our horses. Then I saw a narrow track of firm grass, just wide enough for one horse. I urged my horse onto it and thus was able to pull away from the dragoons, still wading through the mud to my flank. Two dragoons who had lost their horses appeared in front of me. There was nothing for it but to go straight at them. I bowled one of them over and I was through.

A closed line of cavalry was approaching me from the right. I took them for Prussians and rode towards them. Then I saw that they were wearing bearskins so I turned off to the right. After a short while, I met a troop of Kürassiers who had gathered themselves around Premierleutnant von Poser, commander of the Standartenzuges [escort to the standard].

Here the remnants of the regiment gathered, a small group which was quickly organised into troops. We didn't have

The Clash at Liebertwolkwitz

much of a rest. Large groups of French cavalry appeared and came at us. We charged against them, but didn't stand much of a chance against their superior numbers, until suddenly the Neumark Dragoons hit the enemy in the flank. We now joined up with our dragoons and charged the enemy and threw them back into a deep, sunken lane where many men and horses fell and filled the ditch.

We now reformed line and stood at the ready. A heavy artillery duel began in which we were subjected to a lot of shot and shell. Rittmeister von Klöber, our squadron commander, was before the squadron, just by me, and was ordering the ranks, when a shell landed between us, exploded, and a piece broke his stirrup and cut the ball of his foot. The rest of us around him were covered with dirt. His horse, whose girth had been cut, went down.

We thought he was dead, but soon found his wound. We carried him to a covered cart just behind the regiment and he was taken back to Pegau. While he was still near the regiment, a cannonball went through the canvas roof of the cart and the Saxon driver at once whipped up his horses to get out of the danger zone.

Meanwhile, the other regiments of our corps had been involved in some splendid charges against the enemy. The fight raged back and forth. This was definitely one of the most bloody and interesting cavalry combats in military history. Our Neumark Dragoons were very distinguished. Our cavalry charges were so violent that we drove deep into the enemy cavalry formation. The King of Naples was at the point of being captured in one of the mêlées. It is not certain if it was Leutnant von Lippe of the Neumark Dragoons or Major von Bredow of the Brandenburg Kürassiers who challenged Murat to surrender. Some say that the man wore a helmet, in which case it was Bredow; others say that it was Leutnant von Lippe of the dragoons.[1]

In the chase after the beaten enemy, the latter came up on the left side of the King of Naples and shouted: "Surrender, king!" At this moment, the king's master of horse, who

---

1. Prussian dragoons wore shakos in 1813. The sources differ regarding the identity of the officer who nearly captured Murat. Odeleben states that it was probably von Lippe, but that Major von Waldow, Rittmeister von Waldow, and Leutnant B. Richthofen of the same regiment (1st Neumark Dragoons) were also killed there. Zelle states that it was Major von Bredow.

reported the incident, stabbed the brave officer in the side with a stiletto and killed him.

The result of the bloody combat was that our 30 squadrons held the field against an enemy three times as strong and that the French withdrew as darkness fell. We had achieved the aim of the reconnaissance and now withdrew into a comfortable bivouac behind the Universitätsholz between Güldengossa and Störmthal.

The regiment had bought its glory at a very heavy price. Fourteen officers were dead, wounded or missing, over half of the total. About 200 men and horses were out of action. These figures show how much the officers had exposed themselves.

The regiment's heavy losses were due to the fact that it had followed the beaten foe too closely and too far and had left the other Allied regiments too far behind. We also became disorganised around the captured battery and got involved in too many individual combats.

The French took advantage of these circumstances, and of their superior numbers, and attacked us in the left flank with two regiments of dragoons so that the regiment had to fight its way out in small groups, which cost us heavy casualties.

I was one of the last to rejoin the regiment, and several Kürassiers had reported seeing me in the midst of the French and had reported me as dead or captured. They had given me up for lost, and my faithful old Czekyra had shed a tear or two over me. His joy, when he saw me on my exhausted chestnut was thus the greater.'

Sources

Aster, Heinrich, *Die Gefechte und Schlachte bei Leipzig im Oktober 1813*, Dresden, 1856

Odeleben, Otto von, *Napoleons Feldzug in Sachsen im Jahr 1813*, Dresden, 1816

Pflug-Harttung, Julius von, *Befreiungsjahr 1813*, Berlin, 1913

Smith, Digby, *1813 Leipzig: Napoleon and the Battle of the Nations*, London, 2001

Zelle, W. *1813 Preussens Völkerfrühling*, 2 vols., Leipzig, no date

Chapter 11
The Battle of Möckern
16 October 1813

This action, perhaps more than any other in the Wars of Liberation of 1813–15, was to prove that the rotten Prussian Army of Jena and Auerstädt was truly a thing of the past. The achievements of Yorck's corps in taking this village against bitter resistance cannot be too highly praised. 'Möckern' is truly a battle honour that any army could be proud of. Such a feat of arms could only have been carried out by men of all ranks whose morale was excellent. Möckern provides another example of cavalry delivering the well-timed decisive blow to sorely-tried infantry in square, and breaking that square to seal the victory.

In contrast to Schwarzenberg's confused fumblings in the south of the developing Leipzig battlefield, Blücher, as commander of the Army of Silesia, had one very clear aim: to close with Napoleon and destroy him. It was to this end that he had made his bold advance westwards out of Silesia, through the Oberlausitz and down the Elbe to Wartenburg, following Bülow's defeat of Oudinot's thrust on Berlin at Gross-Beeren on 23 August and his own victory over Macdonald at the Katzbach on 26 August. He had forced the crossing of the Elbe at Wartenburg on 3 October and headed south-west, turning Napoleon's northern flank.

Leaving Tauentzien's IV (Prussian) Corps in the Dessau area to guard the approaches to Berlin, Blücher then swung around the enemy's flank – dragging the unwilling Bernadotte and his Army of the North along with him – in a manoeuvre reminiscent

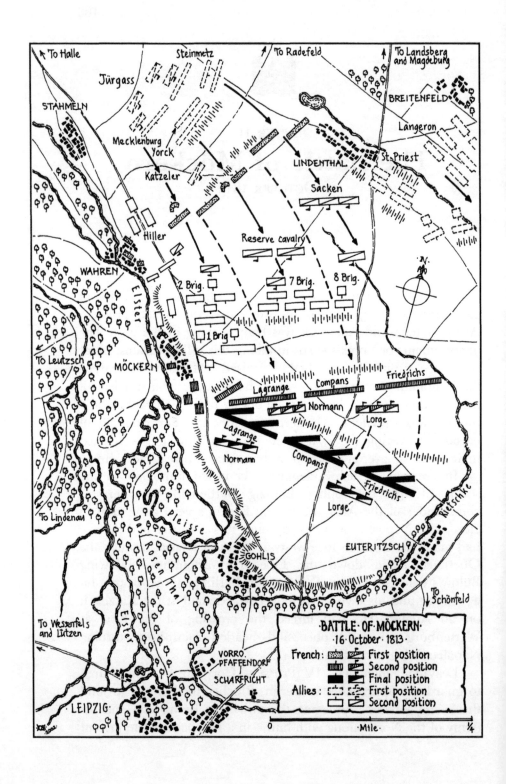

The Battle of Möckern

of Napoleon at Ulm in 1805. As the two Allied armies closed in on Leipzig from the north, the Army of Bohemia approached from the south. The conjunction of the Allied forces, which Napoleon had to prevent, was in the process of completion. To block it, the Emperor left Dresden on 7 October, intending to destroy Blücher in isolation. Forewarned of this advance, Blücher crossed to the west bank of the Mulde on the 10th, and by the evening of the 12th was on the west bank of the Saale at Halle, whilst Napoleon – with the Imperial Guard – was far to the east at Düben on the Mulde.

Hampered by his lack of scouting cavalry, the Emperor had lost the initiative in the campaign and had completely lost sight of the armies of Bernadotte and – worse – Blücher. He was about to receive a very unpleasant surprise. Langeron's corps was the first of Blücher's formations to close with Marmont's troops at Radefeld, north of Möckern on 16 October. Even more fortunately for the allies, the Emperor had reconsidered his dispositions around Leipzig during the night of the 15/16 October. Convinced that there was little serious immediate threat to his northern flank, he decided to transfer some of the troops from there to stiffen up the southern side of his defensive ring. Marmont was ordered to evacuate the lines that had been dug and parts of his force were already marching off to reinforce Napoleon's southern flank. This left about 34,300 French and allied troops facing the Army of Silesia. There were 18,600 men in Marmont's VI Corps, which was supported by the cavalry divisions of Lorge and Fournier.

Langeron's Russians were the first to close with the enemy; Yorck and Sacken followed him. At about 2 o'clock in the afternoon, Yorck ordered the assault on Möckern village to begin. The garrison of the place was made up of the 2nd and 4th Naval Artillery Regiments, a misnomer as these seamen were being used as infantry. The village had been well prepared for defence and the assaults were repeatedly beaten off with loss. Artillery shells soon set fire to Möckern and this added to the terror and confusion of the scene. There was house-to-house fighting through the blazing streets and buildings and still the French held on. By late afternoon the fate of the day still hung in the balance. The contesting infantry surged back and forth in a confused mass in front of a French battery at the town's cemetery. The decisive moment was near.

Yorck rode forward to Major von Sohr, who was behind the Prussian infantry with the Brandenburg Hussars, and shouted to him through the din and the smoke: 'Charge! If the cavalry don't do something now, everything is lost!' Sohr pointed out that his three squadrons alone were incapable of bringing about the desired decision. Yorck realised he was right and sent an adjutant back to the Reserve Cavalry. After ten minutes (which seemed like years to those waiting) they trotted up. Sohr gave the order: 'Trumpeter, sound the trot!' and his regiment moved forward against the advancing French line. The shock of the cavalry threw the startled enemy infantry into confusion, and they were overthrown and cut down. The survivors fled back onto – and through – the battery; six guns fell into Prussian hands.

The Prussian cavalry had thus turned the day. The usually terse and laconic Yorck met the wounded Major von Sohr (holding his sabre in his left hand after having been shot in the right arm) after the action was over. Yorck told him: 'I owe today's victory to you alone and I will never forget you and your regiment.'

Graf Normann rushed up with his 25th (Württemberg) Cavalry Brigade to try and restore the situation for the French, but just as he was about to charge into the flank of the Brandenburg Hussars he was himself taken in flank by Oberst von Katzeler at the head of the Brandenburg Ulans and the 5th Silesian Landwehr Cavalry. The Württembergers were overthrown and driven back into the French infantry behind them, who broke in their turn. The Brandenburg Hussars also cut into them, taking more prisoners and another gun.

The seven Prussian reserve cavalry regiments now thundered along the French line hacking and stabbing at everything in their way. The crisis had passed. The French were crumbling and breaking away to the rear. Just one more push was needed. Yorck therefore called up his last cavalry reserve, the Mecklenburg-Strelitz Hussars, and urged them on into the melée.

The Mecklenburg-Strelitz Hussars had, as usual, been given the task of covering their artillery battery. While waiting

The Battle of Möckern

patiently through the long hours of enemy artillery fire that this entailed, another cavalry regiment had ridden up behind them in full parade dress, with long white plumes in their hats and with drawn sabres. They were the Swedish Mörner Hussars from the advanced guard of the Army of the North, who had come up to see the enemy. The main body of their army, they said, was a day's march behind them. They declined an invitation to join in the battle and rode off again. Shortly after the Swedes had left, the British General Sir Charles Stewart (liaison officer to Bernadotte) also rode up, resplendent in his red uniform, cocked hat and plume, and mounted on a magnificent horse. He rode quietly past the Mecklenburgers and directly towards the enemy, as if out on a quiet hack in the English countryside. A cannonball, which only just cleared his head, caused him to do a smart turnabout and make off for a safer spot at a speed reminiscent of Ascot racecourse.

Yorck's order for the Mecklenburg Hussars to attack came at about 5 o'clock, and the regiment moved thankfully forward, just as the wounded Prinz Karl von Mecklenburg was carried past them to the rear. Sickness and the provision of a detachment of 150 men for duty elsewhere meant that regimental strength was down to just 280 sabres. As they rode past General von Yorck, he indicated the enemy infantry line to their commander, Oberstleutnant von Warburg. 'They've been near to breaking once already. They're steady now, but if they start to waver, charge them!'

Finally, the moment came. As the French infantry fell back, the Mecklenburgers charged. The French infantry formed square but the hussars broke it at once and the slaughter began. A cry of 'Enemy cavalry!' caused the Mecklenburgers to rally outside the square to meet this threat, but Ordonnanzoffizier Timm saw a French officer running off with an eagle. He raced after him, cut him down and seized the prize.[1] The rest of the enemy infantry, seeing the eagle taken and Prussian infantry coming up, surrendered. An immediate count of prisoners realised one colonel, two lieutenant-colonels, 21 other officers and 384 NCOs and men.

---

1. The history of the Mecklenburg Hussars states, wrongly, that the eagle was that of the Sailors of the Guard.

The Battle of Möckern

Now Marmont's troops lost heart. Everywhere his infantry squares were bursting apart or dissolving to the rear as Prussian cavalry swirled among them. The diary of the Lithuanian Dragoons described the dramatic scene as follows:

> 'We had broken through the enemy mass. Regardless of the hail of lead, we had penetrated into the centre of the thickest formation when suddenly the defeated French behind us regained their fighting spirit, picked up the weapons they had just discarded, and fired into us again. The result was a gruesome bloodbath. It was a dreadful scene. We pushed on into the terrified crowd. Those that were not cut down by our sabres were trampled down by the horses. The unfortunates lay in heaps of 20–30.
>
> Certainly, not one of this regiment of Sailors of the Guard[2] would have survived had we not suddenly received a discharge of artillery fire from the left which caused us to fall back. Nevertheless, we had taken most of them and those who escaped left their packs and muskets in our hands so that they could run faster.'

In the course of that evening the Mecklenburgers brought in over 500 more prisoners, and a howitzer, complete with team. Of the regiment's own 18 officers, three were killed and three wounded. Werner Behm served in the Mecklenburg Hussars in this action and left us his account of the breaking of the square:

> 'At last the long-awaited moment came. Major von Schack of Yorck's staff rode to a small hill to the front left of the regiment, where a clearing in the smoke allowed him a better field of view. He waved his hand. The regiment moved left behind Captain Huet's battery and went into a trot; the cannon followed. We could clearly see the enemy in the act of retiring. To the right we saw the flashes of their guns and hoped that we would be out of their range.

---

2. As has already been pointed out, the Marines de la Garde were not present. The unit which was cut down here and lost its eagle to the Mecklenburg-Strelitz Hussars was the 1er Régiment d'Artillerie de la Marine (the heroes of Lützen and Dresden), of which 33 officers were killed during the fight. Martinien shows the Marine Artillery as being at Wachau on 16 October, but this is wrong, as is Zelle's attribution of the capture of its eagle to the Prussian Lithuanian Dragoons.

The Battle of Möckern

Already we could see an enemy square. Both wings of our regiment extended beyond it, but all veered in to the centre and a dreadful crush developed, so that in the 2nd Squadron, some horses and riders were lifted off the ground. Luckily for us, the side of the square facing us fired too soon. The order: "March! March! Hurrah!" rang out and we charged at the enemy.

A group of riders in advance of our main body crashed into the corner of the square like a battering ram and broke it away from the other infantry. Oberstleutnant Warburg ordered the wings of the regiment to close in and we surrounded the square on all sides.

On the left wing Leutnant Schüssler was killed and Rittmeister Damm was wounded in the arm; on the right wing Leutnant von Hobe and Major von Bismark fell from their horses badly wounded.

At the first shock, several of our men had penetrated the square in various places, including Gefreiter Woltersdorf, whose horse received several bayonet wounds in the chest, but crashed into the square and out again.

At the front of the square, which had been the steadiest side, Gefreiter Benzien of the 3rd Squadron had charged into the thickest part and had caused chaos, as had Gefreiter Rheinhold of the 4th Squadron. Many other troopers followed their example. We later counted 60 horses with bayonet wounds in the chest.

In a few places the square held firm. At these points, the troopers turned their horses to the left in order to be able to use their sabres, and edged up to the enemy, trying to hack at the officers who were ordering their men to fire. These suffered many cuts to their faces.

Gradually the square was crushed together into a shapeless mass; there was no way out for them. The hussars shouted "Jettken in Arm!", which was as close as they could get to "Jettez les armes!", but the enemy refused. Many duels broke out between hussars and the infantry.

Rittmeister von Lüttichau, commander of the Freiwilligen Jäger Squadron, spurred after four French officers who were trying to escape and ordered them to surrender. They stopped and approached him, making as though to give up their swords. As he bent to accept them they sprang at him, snatched his sabre, grabbed the reins of his horse and shouted: "Vous êtes notre prisonnier!"

Some of his Jägers saw what was happening and rushed to his aid and brought the four officers back to captivity.

Gefreiter Lange and Jäger Victor von Örtzen caught up with a colonel who was trying to escape; he surrendered his sabre without a fight.

Oberstleutnant Warburg, who had been between the 2nd and 3rd Squadrons, had broken into the square followed by Leutnant von Kamptz and several troopers, and confronted a group of French officers, calling on them to surrender. He was answered with a sword cut to the left hand; in response Warburg hacked his opponent on the epaulette, cutting him down.

Then we heard the call: "Enemy cavalry!" Warburg gave orders for all available men to concentrate to left and right of the square. Trooper Timm was left alone in the square. He saw two French officers trying to escape and rode after them. He rode one down and cut the other with his sabre. As he did so, he saw a gilt eagle sticking out of his coat. "That's the cuckoo!" shouted Timm, and cut the officer down. Timm dismounted to claim his prize but the dying ensign fought him off and held on to his treasure until Timm hacked off his hand.

Now the eagle was his, and it was an eagle of the Imperial Guard! "Look at the cuckoo!" shouted Timm as he rode to the colonel.

That same evening, Timm was sent to Blücher's headquarters with the eagle. Next day, Blücher sent it on to Alexander's headquarters.

Timm was rewarded with Prussian and Russian orders and awards. As he watched the trophy being passed from hand to hand by the generals, he said: "Now that I've tamed the bird of prey, it's easy enough to make it hop from finger to finger; yesterday you wouldn't have been so keen to play with it!"'

A French account of the destruction of the 1st Naval Artillery Regiment is taken from the memoirs of Captain Jean-Louis Rieu, a company commander:

'We were deployed in line as if for a review, but in two ranks

The Battle of Möckern

instead of three so as to show a longer front, and this was a bad sign. I can still hear Mutel, who carried the colours, asking whether he shouldn't put the eagle back in its case because its glitter in the sunshine provided the enemy with a target; and the major's reply, at the top of his voice, that on such a day one could never let an imperial eagle shine too much. Very soon the profound calm was succeeded by the din of artillery and musketry fire. Our only battery was smashed in the twinkling of an eye by the enemy's formidable artillery, and as a crowning disaster, an ammunition waggon full of shells and charges caught fire and spewed death on all sides. Our skirmishers were pushed in by superior numbers. In readiness for receiving cavalry we changed from line into mass formation, but the grapeshot merely ploughed deeper gaps in our ranks. However, we stood fast in the hope that reserves would arrive to support us – vain hope!

There we were, suffering increasing damage from the grapeshot, still in line by battalions in mass formation. No orders reached us. We could hear no leader's word of command, and we felt that we had been abandoned on the battlefield. This is explained by the fact that Marshal Marmont and General Compans had been wounded. I am not sure whether General Pelleport had been, too, but in any case I didn't see him again. As for my blustering little major, he was nowhere to be seen! I discovered later that he had used a scratch as a pretext for retiring shamefully from the mêlée, and so had the lieutenant in my company, both of them without saying a word to anyone. If the major in command of the regiment gave no sign of his presence, it is no doubt because he was bewildered by the turmoil. At least he did not run away, and we met up with him again as a prisoner.

Meanwhile the Prussian infantry battalions were approaching so close, thanks to their artillery support and our immobility, that their ranks met ours, so much so that a sergeant-major named Mourgue took them to be French on account of their blue greatcoats, which were like our own. He went unofficially to one of these battalions to warn them that they were firing into their comrades and was very lucky to escape being captured.

Our position was becoming untenable. Besides the enemy

artillery which was killing us at point-blank range, an imposing force of cavalry waited a mere 20 metres away for us to break, when they would spring at us like tigers waiting for their prey. Our companies were becoming more and more disorganised, and very soon the battalions, being crowded together, presented nothing more than unformed heaps, which still fired a few shots and whose officers no longer had any influence unless they stayed there in person and physically held the soldiers back. This could not last very much longer. The instinct for self-preservation, even if prompted by bad motivation in these circumstances, became too strong. The men broke and fled.'

The report of Oberst von Wahlen-Jürgass, commander of Yorck's cavalry, described the action from his perspective:

'At about 1 o'clock our infantry had finally been able to penetrate into the village and the commander [Yorck] ordered the cavalry to advance and assault the enemy left wing. I carried out this order at once by leading the 1st West Prussian Dragoons and the Neumark Landwehr Cavalry Regiment forward at a trot with the Lithuanian Dragoons following in support. The West Prussians hit the enemy [some squadrons of Chasseurs] first, overthrew them and took four limbered cannon, then chased the enemy, together with the rest of the cavalry, back on Gohlis, taking several hundred prisoners and cutting down many more along the way. Then I saw several intact columns of enemy infantry off our left flank. I left the West Prussians in front of Gohlis, where the enemy had reformed. I then gathered in the rest of the brigade which had become disordered in the chase as is usually the case. The Lithuanians were still in compact order. I ordered them to wheel about and to charge the nearest square. This they did and cut them down where they stood as a later inspection of the battlefield showed. The enemy infantry to left and right of this square scattered. Those directly behind it fled to the rear in such a dense mob that the cavalry could not force their way in. At last we gave up the chase because of the dark.'

Sources

Andolenko, General C.R., *Aigles de Napoleon contre les Drapeaux du Tsar*, Paris, 1969

Behm, Werner, *Die Mecklenburger 1813 bis 15 in den Befreiungskriegen*, ten volumes, Hamburg, 1913

Bodart, Gaston, *Militär-historisches Kriegs-lexikon*, Vienna and Leipzig, 1908

Chandler, David, *The Campaigns of Napoleon*, London, 1966

Droysen, Johann Gustav, *Das Leben des Feldmarschalls Graf Yorck von Wartenburg*, Leipzig, 1890

Esposito, V.J., and Elting, J.R., *An Atlas of the Napoleonic Wars*, London, 1999

Martinien, A., *Tableaux pars Corps et pars Batailles des Officiers Tués et Blessés Pendant les Guerres de l'Empire 1805–1815*, Paris, 1890

Seyfert, Friedrich, *Die Völkerschlacht bei Leipzig vom 14. bis 19. Oktober 1813*, Dresden, 1913

Six, Georges, *Dictionnaire Biographique des Generaux & Amiraux Français de la Revolution et de l'Empire (1792–1814)*, Paris, 1934

Zelle, W., *1813 Preussens-Völkerfrühling*, 2 vols., Leipzig, no date

Chapter 12

Allied Cavalry Raids of 1813
March–October 1813

Reference has already been made to the use of cavalry raids deep into enemy lines of communication. Some generals did not think these worthwhile. Napoleon for one did not favour this tactic. He preferred to use his cavalry *en masse* to deliver knock-out blows on his chosen battlefields. He clearly thought that raiding operations dissipated assets to little effect. The Russians, however, made extensive use of this tactic in the 1812 campaign, seemingly influenced by reports of the success of the Spanish guerrillas in French rear areas from 1809 onwards. However effective such operations might or might not be, they were certainly an assignment that was best given to light cavalry or irregulars such as Cossacks, who were less appropriate for major shock action. This chapter will look at what such raids achieved in the campaign in Germany in 1813.

By 1813 Napoleon's hold over Germany was loosening. The relentless imposition of his Continental System since 1807 had bankrupted much of the traditional commerce of the continent and led to a flourishing smuggling business as British goods flooded into Europe despite all that he could do to stop them. The imposition of the Continental System had led directly to the annexation of the Kingdom of Holland, the Hanseatic cities and all the non-Prussian Baltic coastal regions in Germany. Prussia itself was awash with French agents and had been kept under Napoleon's rule in all but name. However, by 1813 the mighty, politically monolithic, but economically unsustainable edifice of

Allied Cavalry Raids of 1813

the Napoleonic empire was beginning to crumble as the allies advanced from the east. The upsurge of Prussian nationalism was also infecting the unhappy populace in the puppet grand duchies and kingdoms in western Germany. Indeed, the climate in Germany was beginning to mirror that in Spain, where guerrilla groups could find friendly waters in which to swim, usually secure from capture by the forces of an unpopular state.

Under Russian influence, the allies agreed to contribute limited resources to form a number of *Streifkorps* (raiding parties), which were to carry out deep penetration raids behind enemy lines. In effect, they were forerunners of the Long Range Desert Group and the Special Air Service of the Second World War, and the many other similar organisations which have since been raised. Of course, the raiding parties of 1813 were not tasked to take out any particular strategically important enemy command organisation; they were in no sense the 'Scud Busters' of the Gulf War. Their job was to go behind enemy lines and create as much mayhem and chaos as possible, on such opportunity targets as presented themselves. Their composition was mainly cavalry, with a small element of infantry and some horse artillery guns. The three combat arms were thus present, but the proportions of each showed very clearly that their tasks were to hit and run, not to occupy the areas in which they operated. Despite this, some of these forces found themselves in possession of major towns and cities for significant periods.

Even prior to the temporary armistice agreed between Napoleon and his enemies in the summer of 1813, allied raiding parties had, for example, occupied Hamburg – which was then part of France – and were pushing into the Harz Mountains, deep into the Kingdom of Westphalia. The terms of the armistice, agreed between the warring parties in June, stated that all allied forces would vacate the left bank of the River Elbe by the 12th of that month. This was all well and good, but no-one had thought about how to communicate this arrangement to the partisan groups, still deep in enemy territory, at unknown locations.

That these groups were a considerable thorn in Napoleon's side may be judged by the fact that, during the armistice, he tried to destroy Lützow's *Freikorps* at Kitzen in direct contravention of the terms of the armistice. With truly Macchiavelian talent, he ensured that another German force (Württembergers) would be used for the job, obviously seeking to drive wedges into the emergent alliance, which he saw developing. Other allied *Freikorps* had similar problems. Colomb's force was attacked by the Westphalian General Hammerstein at Köthen on 22 June, but managed to escape with the loss of 14 men.

Some of these small, international corps were *ad hoc* groupings of a few squadrons and companies, which were dissolved at the end of the campaign. Others, raised by patriotic action in Prussia, took on lives of their own and were later converted into regular regiments. A listing of the major groupings appears in the appendices (see pages 292–93), but some raids were also made by allied advance guard cavalry.

We will now review the raids of 1813.

Hamburg, March–May 1813

The city of Hamburg lies on the estuary of the great River Elbe, which flows into the North Sea. In 1810 it had been annexed into metropolitan France by Napoleon as part of his increasingly desperate effort to stem the large-scale smuggling of British goods. The 127th Line was raised on 1 March 1811 in the city and in Bremen. The population of the old Hanseatic city had been ruined by Napoleon's Continental System, which had reduced the once-bustling harbour to a stagnant backwater. In January 1813 it was occupied by General Carra St. Cyr and the 1st Division of the Corps of Observation of the Elbe.

Word of the destruction of the Grande Armée in Russia and of the advance of the allies raised hopes that the hated French would soon be expelled, and the rumblings of rebellion soon reached St. Cyr's ears. He knew that he had no chance of holding down the city with his small force, so he evacuated Hamburg on 12 March, and retired to the south bank of the Elbe. As allied forces approached, he withdrew south-west into Bremen on 21 March.

To make matters worse for Napoleon, General Baron Joseph Morand evacuated Mecklenburg and fell back south-west to join St. Cyr in Bremen on 22 March.

Allied Cavalry Raids of 1813

The Emperor was livid. These timid actions had torn a great breach in his coastal 'Western Wall', and the Baltic and the North Sea coasts were open. The hated British trade, together with troops, weapons and supplies would pour in. St. Cyr was recalled in disgrace and placed under Vandamme's command.

Into this turbulent power vacuum rode the Russian Colonel Tettenborn with a Cossack *pulk*, 1,300 men strong. The townsfolk invited him in and he agreed to enter, but only on condition that the city fathers return to the pre-French constitution. This they promised and on 17 March a group of 12 Cossacks rode formally through the city's gates. They were greeted as heroes by the cheering crowds and, in no time at all, had been plied with so much drink that all were laid out in the road. It was not until 30 May that France's iron fist, in the form of Marshal Davout, closed around the city again. The retribution he exacted from the unfortunate populace led to him being dubbed 'Davout the Terrible'. Even so, for an incredible two and a half months, by invitation of its 'French' populace, a handful of Cossacks had kept control of this great city.

The Ambush at Wrietzen, 17 February
This small town is situated about 30 miles north-east of Berlin. The Russians sent a Cossack *pulk* of 500 men under Lieutenant-Colonel von Benkendorf to sweep the area; they surprised and captured a complete battalion of Westphalian troops under Colonel von Seyboldsdorff, without a fight.

Berlin, 20 February
On 20 February Czernichev rode into the Prussian capital on the one side as the French garrison marched of the other. The populace went wild with joy.

Dresden, 10 March
Davidov's *Streifkorps* and a squadron of the 1st Bug Cossacks entered into a capitulation agreement with the city authorities whereby the Russians took possession of the new part of the city. For his pains Davidov was sacked by Winzingerode and sent off to Kutuzov to be court-martialled for concluding a truce with the enemy without permission.

Langensalza, 16/17 April

This town is in the Thuringian hills, north-west of what was then the French-held fortress of Erfurt, hundreds of miles behind the lines. Major von Hellwig's Prussian *Streifkorps* of 150 men of the 2nd Silesian Hussars ambushed a Bavarian brigade of 1,551 men and a battery of artillery under General-Major Count von Rechberg. The Bavarians lost 45 killed and wounded, 12 captured, five of their six guns, and three ammunition wagons; von Hellwig's losses were very slight.

Wanfried, 18 April

Wanfried lies about 25 miles west of Langensalza; von Hellwig struck again, this time at the Westphalians. His prey was Oberstleutnant von Göcking, with a company of infantry and a squadron of hussars, some 132 men in all. The morale of the Westphalians was rock bottom. They just dropped their weapons. Three officers and 103 men were captured.

Von Göcking had previously been in Prussian service, prior to 1807. Now he applied to be readmitted, and in 1815 served as a volunteer. He later became a major in the Prussian Landwehr. Many of his men also took service with the Prussians in the days following the raid.

On 23 April 1813 von Hellwig was ordered by King Friedrich Wilhelm III to raise a corps to act in a partisan role in the French rear. Blücher added that he 'might go wherever, whenever he wished'. After the end of the armistice, Hellwig joined von Bülow's corps and fought with it through Germany, Holland and Belgium. The corps fought at Bautzen on 25 May 1813, Schoenweide on 20 August, Grossbeeren on 23 August, Dahmsdorf on 24 August, Speerenberg on 25 August, Holzdorf on 6 September, Wartenburg on 21 September, Wustwetzel on 7 and 11 January 1814, Peer on 20 January, Lowenzout on 28 January, Tournai on 8 February and Ypres on 23 February.

Nordhausen, 19 April

The town lies about 50 miles west of Halle in Saxony. Russian Major-General Landskoi and his Cossack *pulk* surprised a squadron of Westphalian hussars of 106 men; they surrendered without a fight.

Halberstadt, 30 May

Halberstadt lies about 55 miles north-west of Halle in Saxony. General Czernichev, with the Riga Dragoons, the Isum Hussars and four Cossack *pulks*, some 1,200 men and two guns, surprised the Westphalian General de Division von Ochs in the walled town with a motley force of 1,200 men (recruits, invalids, veterans and gendarmerie) with 14 guns. General von Ochs's force lost 33 killed and 40 wounded. The general, eight officers, over 1,000 men, 16 guns and 80 ammunition wagons were captured for the cost of 43 killed and wounded.

Kitzen, 17 June

Kitzen is a small town about nine miles south-west of Leipzig. Colonel Baron von Lützow had been operating with his Royal Prussian *Freikorps* west of the city prior to the conclusion of the armistice and was ignorant of it. He had been rampaging through the countryside, robbing government treasuries, scattering troop transports, distributing leaflets in support of the allied cause and recruiting volunteers and deserters. In Halle, Saxony, scores of students flocked to join his party. On 9 June Lützow was informed of the armistice by the Bavarian authorities in Hof and of the fact that he should be over the Elbe by 12 June. He did not trust the Bavarians, so moved to Plauen and asked the Saxons about it. Only on 14 June did the Saxon Ministry of War confirm the news to him. Lützow decided to cross the Elbe at the closest point – Leipzig.

On the evening of 16 June, he wanted to bivouac at Zeitz, but found it occupied by a force of Württemberg troops under Oberst von Kechler. He aimed to go around the town to Kitzen, sending Kechler a message explaining his intentions, and waited for a reply.

That same day, the Württemberg brigades of Döring and Normann were in Leipzig; they had received the following orders from General Arrighi, the Duke of Padua:

> 'On the Emperor's orders, Colonel Prince Hohenlohe will take the 1st Battalion of the 4th Infantry Regiment and two squadrons to occupy that part of the Duchy of Dessau lying on the right bank of the Elbe. At the same time, four mobile

columns, each of 200 infantry and 100 cavalry will be formed, in order to intercept the partisan groups of Czernichev and Lützow, which are still in the rear areas of the French Army.'

Kechler now reported the presence and peaceful intentions of Lützow's troops to Arrighi who sent a message that Kechler should tell Lützow not to move, but to await the arrival of a conducting officer that Arrighi would send. Kechler and Lützow accordingly met. Kechler delivered this message and added that Lützow would run into trouble if he moved off, as several columns were out looking for him. Lützow agreed to wait, but requested a Württemberg officer as hostage as he did not trust the French. Kechler refused, but gave him his word of honour that he would not undertake any hostile actions. The two officers shook hands, and Lützow agreed to await the conducting officer.

Kechler then learned that Graf von Normann's column was on the way from Lützen to Kitzen. Kechler galloped to meet Normann and explained the situation. With Normann was the French General Fournier, whose response to the news was: 'Attaquez les Prussiens!' He added to Kechler, that he would have his head if one Prussian escaped. Kechler responded that he had met with Lützow half an hour before and told him of Arrighi's aim. At first Fournier was very angry, but after a while, he called Kechler to him and said that he had done the right thing.

Kechler went back to his column, which he did not find until dark. When he arrived he at once sent an NCO to Lützow to warn him that he was under orders to attack him, but the NCO failed to locate the Prussians. In the meantime Kechler set off, marched for some hours, then heard shooting and made towards the noise. At about midnight, he found General Fournier at the village of Krautnauendorf. Arrighi had heard of the presence of the Prussians from another source, prior to Kechler's report and was determined to carry out Napoleon's express order to catch and destroy them.

Fournier commanded a battalion of naval infantry, 200 French dragoons and Graf von Normann's Württemberg detachment of two companies of infantry, two squadrons of the

Leib-Chevauxlégers, and three guns. Normann was now ordered to occupy Kitzen on 17 June. When he approached the place he found his way barred by five Prussian squadrons. Lützow came forward. Normann told him of his orders and, after a brief conversation, Lützow rode off to speak with Fournier. Normann guaranteed not to attack Lützow's force until he returned.

Whilst waiting for Lützow's return, Normann deployed his men and the Prussians also began to move off towards Leipzig. About ten Prussian officers were with Normann, when a French orderly rode up from Fournier and said to Normann: 'Le general vous fait ordonner, de faire arrêter a son retour le colonel, qui parle dans ce moment avec lui.' ('The general orders you to arrest the colonel to whom he is now speaking when he returns.') Apparently none of the Prussians present spoke French, so did not understand what was going on. The plot was clear, though. Fournier wanted Lützow's head, but he was looking for other Germans to do the dirty work.

Suddenly Lützow came galloping back, joined his cavalry and they trotted off. Moments later, Fournier rushed up and ordered Normann to catch them – the chase was on! It was now late in the day and getting very dark. The Prussians halted by a village, probably Klein-Schkorlop, and deployed for combat. There was a clash, and the Prussians scattered into the darkness. Since night had fallen Fournier broke off the action and the troops went into bivouac.

The Württembergers lost one killed and three wounded; the Prussians lost ten officers, 100 men and 65 horses captured. Fournier was furious that Normann had not carried out his instructions, but Normann answered that he had given his word of honour to Lützow that he would not attack him, and anyway, Fournier could have arrested him if he had so wished. Arrighi also accused Normann of deliberately letting the Prussians escape. The unfortunate Normann was in a no-win situation, and it was soon to get much worse.

On 18 October, on the field of the battle of Leipzig, he took his brigade over to the allies. For his king, Friedrich of Württemberg, one of the most vindictive and merciless monarchs of the era, this was the last straw. After Leipzig, he gave orders that Normann and Oberst von Moltke, commander of the Jägerregiment-zu-Pferde König, were to be arrested as soon as they re-entered his kingdom. Both cavalry regiments of Normann's brigade were disbanded and the officers and men all lost all their medals and awards. Luckily, Normann and Moltke were warned of their impending fate and fled to Austria. Moltke later became a Feldmarschall-leutnant in Austrian service, but no post was found for Normann. It was not until after the king's death on 30 October 1816 that Normann was permitted to re-enter his homeland and, even then, he was banned from Stuttgart. Normann eventually died of typhus on 15 November 1822 in Missolonghi, fighting for Greek independence.

Brunswick, 25 September
This city lies in northern Germany, 25 miles east of Hanover; in 1813 it was part of the Kingdom of Westphalia. It was held by the aged General de Brigade von Klösterlein, with a company of the Jäger-Carabiniers of the Westphalian Guard, the depot troops of the 1st, 3rd, 4th and 9th Infantry regiments, a company of veterans, a company of departmental gendarmes and a detachment of infantry from Lippe and Waldeck, some 430 men in all.

On 25 September Oberst von der Marwitz appeared at Brunswick with 400 men of the 3rd Neumark Landwehr Cavalry Regiment and demanded that Klösterlein surrender the place. Despite his superior numbers, it was clear to the old general that his men would not stand so he left the Jäger-Carabiniers in Brunswick and made off with the rest of the men south to Wolfenbüttel. The Prussians entered the town unopposed; the guards just threw down their weapons. Only at the August Gate was there a slight skirmish, which was quickly over. A troop of Prussians chased after Klösterlein's men, but they were already in Wolfenbüttel. As soon as Klösterlein heard of the fall of Brunswick, he started out for the town of Goslar, further to the south, with his troops. The Prussian cavalry followed him; at the village of Halchter they caught up with him. Klösterlein deployed his men behind a bank, but they just threw down their weapons. Some 25 officers and 350 men were taken; the rest had fled. Many of these Westphalian troops joined the Prussians on the spot.

Allied Cavalry Raids of 1813

Altenburg, 28 September
This town lies 25 miles south of Leipzig, just west of the River Pleisse. In late September 1813 General Lefebvre-Desnouëttes, commanding a large, composite force of French line cavalry and cavalry of the Imperial Guard, was in the town. His command consisted of a squadron of Mamelukes, four of Grenadiers-à-Cheval, six of Chasseurs-à-Cheval and ten of the 2nd Chevaux-légers; from the I Cavalry Corps, under Général de brigade Piré were eight squadrons of the 6th, 7th and 8th Hussars; from the III Cavalry Corps was a squadron of the 5th Chasseurs; from the V Cavalry Corps three squadrons of the 14th Chasseurs and three from the 6th and 19th Dragoons. There were two batteries of horse artillery. The only infantry involved were four companies of the Baden Infantry Regiment Nr 2 Graf Hochberg. The force totalled some 6,500 men with 12 guns. The allied ambush force was made up of the *Streifkorps* of Thielemann and Mensdorff-Poully and part of Ataman Count Platov's Don Cossack corps of the Army of Bohemia, totalling 1,200 men and four guns.

The allies caught their prey completely unprepared and put them to flight. The 8th Hussars were overthrown and fled the field, riding down their own infantry as they went. The rest of the French cavalry made off in great haste, abandoning the battered German infantry at the bridge over the Gersta stream. French losses were 600 killed and wounded, 1,500 captured, five guns and three standards and colours. The allies lost 200 killed and wounded.

Kassel, 27–30 September
This city was the capital of the French puppet kingdom of Westphalia and was hundreds of miles behind the lines in September 1813. The fact that a raid like this could take place at all speaks volumes for the state of affairs in Germany at this time. The raiders came from General Winzingerode's Russian corps of the Army of the North. They were led by General Czernichev and consisted of Major-General Magnus Pahlen's cavalry division with attached Cossacks, some 2,300 men and six guns.

Czernichev appeared before the Leipzig gate of the city on 27 September. A brigade of Westphalian infantry was sent out to drive him off, but was scattered in a moment. Panic broke out in the town. At this point there were 3,060 infantry, 906 cavalry and 336 gunners in Kassel. King Jérôme handed over command to General Allix de Vaux, a French officer in Westphalian service, and fled in great haste with his Garde du Corps, the Grenadiers of the Guard and the Hussars of the Guard. The garrison left in the city consisted of the Jäger Guard battalion, the Chasseurs-Carabiniers of the Guard, the depots of the Chevauxlégers-Lanciers of the Guard, 2nd, 5th, 7th, 8th Line Infantry regiments, the 4th Light Infantry Battalion, the artillery and the gendarmes. These forces totalled 2,060 infantry, 170 cavalry, 336 artillerymen and 34 guns.

Czernichev now learned that the Westphalian General Bastineller was nearby, in the direction of Halberstadt, with a weak brigade. As he lacked artillery to breach the walls of Kassel, the Russian decided to eliminate his weaker enemy first. Bastineller avoided the thrust and withdrew to the north-east, but as he marched his command dwindled at each step of the way. By the time the Cossacks caught up with him, he had only his officers, some NCOs, 80 men and two guns. One quick charge and it was all over.

General von Zandt of the Westphalian Army had been sent from Kassel north-east to Göttingen with the 1st Battalion, 7th Infantry Regiment, a company of the Jäger-Carabiniers, a squadron of the Chevauxlégers of the Guard and a squadron of the Hussars of the Guard. On 28 September he received an order to return at once to Kassel. He set off immediately but during the night of 29/30 September desertion began to take hold among his men. Next day he reached Kassel with only the Hussars of the Guard and about 200 other troops. By this time, the garrison of the city had been reduced to about 850 men by desertion.

Czernichev assaulted the Leipzig gate on 30 September, eagerly supported by a new battalion made up of Westphalian deserters. The company of Jäger-Carabiniers defending the gate went over to the Russians and it was all over in minutes. General Allix capitulated. The losses of the garrison were a few wounded; there was as good as no resistance. The troops were disarmed and allowed to go home. Russian losses were trivial.

The troops who had accompanied Jérôme were no steadier in their duty and deserted in droves. By the time he reached Marburg on 29 September, Jérôme was accompanied by only 180 men. Most of his Garde du Corps had also vanished by the time he reached Wetzlar. Jérôme continued to Koblenz with 80 of his Guard and an escort of French troops.

Czernichev remained in the city only until 3 October, when he left, taking with him 79,000 francs from the royal treasury. On 7 October General Allix re-entered the place with French troops; King Jérôme followed on the 16th. Decrees were published, calling the troops back to their colours but nobody bothered to obey. The mood of the public was so hostile to Jérôme and his French administration, that they dared not arrest anybody. Jérôme stayed on, trying to remove everything of value that was not firmly fixed down, but left for the last time on 26 October, accompanied by many of his senior generals and administrators, who were seconded to his service from France. The artificially constructed state of Westphalia, created by Napoleon in 1807, collapsed.

The Electorate of Hesse-Kassel, abolished to form part of the kingdom, was reborn. When Elector Wilhelm re-entered his capital on 21 November, he was greeted by a sentry presenting arms. The ultra-conservative ruler looked him up and down and said: 'Wo hat er seinen Zopf?' ('Where's his pigtail?') The men of the re-created army were all ordered to wear pigtails as they had done before 1807.

Bremen, 13–15 October
This old Hanseatic city lies on the River Weser near its estuary and about 45 miles south-west of the city of Hamburg, which place was firmly in the grips of Marshal Davout in the autumn of 1813. The garrison of Bremen consisted of the 1st Battalion of the 1st Swiss Infantry Regiment and detachments of French Customs officials under Colonel Thuillier. It totalled 1,100 men with 14 guns. The raiding party from the Army of the North was led by the Russian General-Major von Tettenborn, with 800 Cossacks and the 1st Battalion of Lützow's Royal Prussian Freikorps. They totalled 1,600 men with four guns.

The raiders attacked the Oster Gate of the city on 13

October, but the garrison just managed to shut it in time. Thuillier ordered that a company of Swiss should make a sortie. This was done, but they fell into an ambush and were mostly killed or captured. Tettenborn then shelled the town, setting it on fire in several places, and Thuillier was killed by a shot fired by one of Lützow's Jägers. Major Devallant assumed command. The defenders ran out of ammunition so Devallant capitulated and was allowed to march out with honours of war, accompanied by the joyous cheers of the populace. This major city-port fell with scarcely a whimper, although the 1st Swiss suffered two officers killed and four wounded and a total of 185 other casualties.

Sources

Andolenko, General C.R., *Aigles de Napoleon contre les Drapeaux du Tsar*, Paris, 1969

Bodart, Gaston, *Militär-historisches Kriegs-lexikon*, Vienna and Leipzig, 1908

Chandler, David, *The Campaigns of Napoleon*, London, 1966

Davidov, Denis, *In the Service of the Tsar Against Napoleon*, London, 1999

Esposito, V.J., and Elting, J.R., *An Atlas of the Napoleonic Wars*, London, 1999

Martinien, A., *Tableaux pars Corps et pars Batailles des Officiers Tués et Blessés Pendant les Guerres de l'Empire 1805–1815*, Paris, 1890

Six, Georges, *Dictionnaire Biographique des Generaux & Amiraux Français de la Revolution et de l'Empire (1792–1814)*, Paris, 1934

CHAPTER 13

The Battle of Fère-Champenoise

25 MARCH 1814

As at the start of 1813, the New Year of 1814 saw the Emperor Napoleon facing new crises wherever he looked. As in 1813, his matchless energy, genius and leadership overcame almost all obstacles – at least initially. His great defeat in the epic Leipzig 'Battle of the Nations' had broken his grip on Germany and the Confederation of the Rhine had crumbled to dust before he was halfway back to the River Rhine. The states which had composed this *cordon sanitaire* between France and her enemies were now arrayed against him, their assets were denied him. Holland and Belgium were becoming restive in the north and were falling rapidly under allied control. His ally, Denmark, was now effectively out of the war. Despite these setbacks the coming campaign was to see Napoleon sometimes at his very best as a commander in the field. These flashes of the old genius, flexibility, improvisation and opportunism were, however, too few and too brief to save him from the determined enemies that he had outwitted and humiliated for so long.

Even within France, there was growing unrest, and many were resentful of the increasingly onerous burdens of financing the Emperor's wars with money and with the blood of their sons. Pro-royalist sentiment was still strong in the west and south of the country. Even Napoleon's marshals were flagging. They had been at war for 20 years and he had made them rich and famous, but they had had precious little chance to enjoy their possessions. His Continental System, hated by all of Europe except himself, had been swept away, so that trade with Britain flourished again. British money was again feeding the coalition against Napoleon. British ships flooded into the Baltic bringing arms, uniforms and equipment to Prussia and other allied states. Britain, his most implacable enemy, that nation of shopkeepers, was near to ruining everything that he had built up over the last ten years.

Napoleon was fighting a war on three fronts: in Spain, Italy and on the Rhine. All his pigeons were coming home to roost. On 10 November 1813 Wellington had broken in the French defences on the border with Spain with his victory in the Battle of the Nivelle River. By mid-December he was battering at the gates of Bayonne and Marshal Soult seemed to have no effective answer to the threat that this allied army posed. In north-eastern Spain, Marshal Suchet still clung on in Catalonia, but his position rendered him scarcely relevant to the decisive struggle which was to unfold in northern France. In northern Italy Prince Eugène was holding the line of the Adige with 50,000 troops, against 75,000 men under Bellegarde, but Austrian pressure was mounting there, as in Illyria and in Dalmatia. Napoleon wanted to pull French units out of Italy for use under his own hand, but Eugène convinced him to leave them in place, as he feared that the rest of his army would melt away if they left. To Eugène's south, Marshal Murat, King of Naples, was in the process of betraying Napoleon and joining his enemies.

Napoleon's main aim was to hold on to Paris. It was his power base, the centralised capital of the nation and a major industrial centre; without the city he was lost. The allies gathered along the River Rhine, exhausted by the strains of the war but determined to see it through to the bitter end. By January they were to have 300,000 men in the field. Napoleon had returned to Paris in mid-November 1813 and at once thrown himself into the Herculean tasks of re-building his army and consolidating his rule. His battered army, now reduced to about 80,000 men, with a high proportion of these sick, crossed the Rhine after the clash at Hochheim on 9 November. The fortress of Mainz was still in French hands, but no attempt was made to hold the line of the river. Indeed, large tracts of territory in eastern France were

The Battle of Fère-Champenoise

evacuated without a shot being fired. Thousands of French troops were still locked up in the fortresses along the Elbe, the Vistula and on the Baltic coast. They busied the Russian and Prussian militia and Landwehr, but most of them fell, one by one, into allied hands.

The Empress Marie-Louise, although of Austrian birth, supported her husband most loyally, urging the young men of France to flock to the colours in defence of their country. Many followed her call; they became known as 'les Marie-Louises'. Indeed the most pressing of the Emperor's needs were new cannon-fodder and more horses. He issued edicts calling up 150,000 conscripts of the class of 1815. Customs officers, sailors, forest rangers, gendarmes and veterans were all called or recalled to the colours and a large portion of the Garde Nationale was mobilised. The total thus ordered to be activated for the defence of France was supposedly 936,000 men. The Emperor also drew three divisions of troops away from the Spanish front to reinforce the army facing the Rhine in mid-February.

Learning from the painful experiences of his armies in Spain, Russia and northern Germany, Napoleon also decreed the formation of partisan groups to harass the rear of the advancing allies. This had perhaps the most depressing outcome of any of his plans at this time: it failed almost utterly. Public enthusiasm for his cause had evaporated. The few cases of French civilians taking action against allied troops were limited to eastern France and involved isolated incidents of retaliation against excessive looting. Wellington went to great pains to maintain good relations with the civil authorities in his area of control, and was singularly successful at it.

The allies were aware of these edicts and measures, which caused them to stop and think rather than rushing into France. What they did not know was that the real manpower achievements lagged far behind the targets, that evasion of conscription, and desertion from the ranks were widespread and that combatting these evils absorbed thousands of troops who would otherwise have been available for service at the front or training recruits in the depots.

Napoleon was tremendously active – and not only in re-creating his army. He also launched into frenzied diplomatic activity to win time and to split the allies if at all possible. He released the Pope from house arrest to placate Italian displeasure. He offered to place Ferdinand VII on the Spanish throne to appease the implacable enemies he had so carefully created by his meddling in their affairs since 1808. He made peace overtures to his enemies. All this to no avail as we shall see. Creating some degree of dissent among the allies, however, was all too easy; they often fomented it themselves. Britain's long term aim was to ensure that no single European mainland power emerged as omnipotent but Russia, Prussia and Austria were always jockeying amongst themselves to better their prospects, to regain provinces lost in the last decades and to snuff out emergent republicanism, wherever it sprouted. This political diversity often affected military planning, and in the past had led to disjointed actions in the field, which had then played straight into Napoleon's hands.

The allies still fielded the three armies of the 1813 campaign, those of Bohemia, Silesia and the North, their ranks now swelling with the contingents of the German states against whom they had been fighting only weeks before. They totalled about 250,000 men but, as with the French, many were sick, clothing and equipment was in shreds and the transport system creaked at every joint. Bernadotte had swallowed the Danish province of Holstein and taken Norway. He was too busy in the north to devote much attention to events in France, though in view of his questionable performance in 1813, this was no loss to the allied cause.

The allied high command decided to keep to the principles of the Trachenberg Plan for strategic operations. Any allied army attacked by Napoleon himself would retire, while the other two closed in on his flanks. There was only one problem: Blücher was so obsessed with destroying Napoleon, that he was repeatedly to throw caution to the wind in the coming fighting, and was to reap a whirlwind every time.

Much to Blücher's disgust, a halt in the allied advance was called in late 1813. The old war-horse champed at the bit at this delay. He passed the time reorganising his troops and building pontoons with which to cross the Rhine. After a few short weeks he was given the go-ahead again. On New Year's Day he led his

The Battle of Fère-Champenoise

Army of Silesia across the Rhine, unopposed, at Kaub, Lahnstein and Mannheim, and headed towards Paris. Schwarzenberg had moved first, though. On 21 December he crossed the upper Rhine between Schaffhausen and Basle and then swung to the right, through Switzerland and the Jura mountains. In doing this, he took care to occupy the Alpine passes to Italy, thus closing them to communications between the Emperor and Eugène.

The allied plan was for Schwarzenberg and Blücher to meet up in the area between Metz and Langres on 20 January and to advance on Paris together. Amazingly enough, the considerable mountain barriers of this region were given up by the French without a shot. The first action of significance on this front took place on 24 January at Bar-sur-Aube, where the Army of Bohemia pushed a French force under Marshal Mortier back westwards on Troyes on the upper Seine. At this point, Blücher had crossed the Moselle with 30,000 men, leaving Yorck to mask the fortresses of Luxembourg, Metz and Thionville. On 27 January, the advanced guard of his Russian corps (Sacken's) was checked at St. Dizier by Milhaud's cavalry, and two days later Napoleon entered the fray with a dramatic victory over Blücher at Brienne. The scales tipped heavily in allied favour on 1 February, however, when Blücher, supported by the corps of Wrede and Guylai of the Army of Bohemia defeated the Emperor at the battle of La Rothière, about three miles south of Brienne. The French lost about 5,600 men and 73 guns and withdrew westwards to Troyes on the River Seine. Allied losses were 6,000–7,000 men but the myth of Napoleon's invincibility had been rudely shattered, and that on the soil of France. The morale of the French people plummeted, as did that of his army.

In the retreat to Troyes in a heavy snowstorm, 4,000 of Napoleon's young recruits deserted, and his reception in the city was decidedly chilly. It was now clear to him that he could not defeat the allies as long as they remained united but Blücher's impatience would give him the chances that he was looking for. On 29 January the allies announced that they would open peace negotiations on 3 February with Napoleon's representatives at Châtillon-sur-Seine. This would offer chances for him to confuse and divide his enemies on the diplomatic front as well.

In the heady aftermath of their victory, Blücher convinced

Schwarzenberg that they should go for Paris. For political reasons, Schwarzenberg was less keen to crush Napoleon than was Blücher, but he agreed to the plan. Blücher was to advance through Châlons and Meaux along the River Marne, Schwarzenberg along the Seine. This would mean that their armies would often be 20 miles or more apart, with only Wittgenstein's corps and Seslavin's Cossacks to link them. It was just the sort of plan that Napoleon had wished for them to adopt.

Initially, things seemed to go well. Sacken's corps defeated Molitor's division of Macdonald's XI Corps at La-Ferté-sur-Jouarre on the Marne on 9 February, and Wrede's V Corps gained a slight edge over Bourmont's brigade next day at Nogent-sur-Seine. For his part, Napoleon had aimed to strike at the Army of Bohemia. The realisation, on 6 February, that Blücher was driving down the Marne on Paris, forced him to abandon this plan and to turn to save his capital city. Blücher was ignoring all the lessons of the Trachenberg Plan, rushing blindly towards Paris, his units losing cohesion day by day. On 10 February things began to unravel for the allies when Napoleon pounced on the isolated division of Olsuviev of Langeron's corps of the Army of Silesia and wrecked it at Champaubert. Next day he struck again at Montmirail and heavily defeated Sacken's corps. His third victory came on 12 February at Château-Thierry on the Marne against Yorck's I Prussian Corps, and on 14 February he defeated Blücher at Vauchamps (Étoges) so badly, that the old warrior suffered a nervous breakdown. His chief of staff, Gneisenau, was in effective command of the Army of Silesia for the next week.

This burst of victorious activity on Napoleon's part has gone down in history as his Six Days' Campaign. It was a dazzling performance which taught Blücher a bitter lesson and cost the allies dear. During this *Blitzkrieg* on the Marne, Schwarzenberg, miles away to the south, moved slowly westwards down the Seine. At the same time, Blücher was being knocked away to the east. By 16 February the gap between the allied armies was over 55 miles.

It was about this time that Napoleon heard the news of Murat's defection to the allies. It hit him very hard indeed, but he quickly put it behind him, changed tack in his operations and moved to destroy the cautious Schwarzenberg. He left Marshal Marmont's reinforced VI Corps to watch Blücher and his shaken army. On 17 February Napoleon struck at the Russian II and VI Corps and defeated them at Nangis (Mormant). Next day he dealt with the IV Württemberg Corps and an Austrian brigade at Montereau. This was enough for Schwarzenberg. He fell back south-east on Troyes and ordered Blücher to close up on him at Méry-sur-Seine by 21 February. Blücher did as he was ordered. Once again, Napoleon had seized the initiative. His army was between the allies, and Blücher's was divided from Schwarzenberg by the flooded Seine.

At the same time Augereau's corps of 20,000 men was advancing from the south and on 22 February was at Lons-le-Saunier, about 80 miles south of the allies at Troyes. Schwarzenberg detached the I (Austrian) Corps, part of the II and Moritz Liechtenstein's Light Division to counter this advance. This detachment caused him to abandon any offensive move against the Emperor.

At about this time, the corps of Bülow and Winzingerode entered northern France from Belgium, with 62,000 men. They were in an exposed and isolated position. The allied leaders agreed that Blücher should join them to form a united force of 110,000–130,000 men. On 23 February Blücher turned and set off for Paris again with his 48,000 men, aiming to join up with these forces at Meaux. Schwarzenberg was to pin Napoleon down in the meantime. The Trachenberg plan had been thrown out of the window. Two days later Blücher was at La Ferté on the Marne, quite isolated from any other allied formation. He was opposed by Marshals Marmont and Mortier with some 10,300 troops. By this time, Napoleon had re-occupied Troyes unopposed while Schwarzenberg fell back behind the River Aube. Feeling suddenly omnipotent again, Napoleon offered peace to his foes, but based on France's 'natural frontiers' of the Rhine, the Alps and the Pyrenees. This was rejected. A counter-offer of peace with France reduced to her borders of 1791, made on 9 March, was thrown out in turn.

Leaving Macdonald, Gérard and Oudinot to mask Schwarzenberg, Napoleon set off after Blücher on 25 February, to destroy him in isolation. On 27 February, Blücher, with Yorck and Kleist, crossed the River Marne at La Ferté to close up to Bülow. Langeron and Sacken stayed south of the river near Château-Thierry. Marmont and Mortier attacked Yorck and Kleist at Gue-à-Tremes the next day, and defeated them in a clash; the allies fell back to La Ferté. By 1 March Blücher's four corps were all around Château-Thierry. Here he heard that Napoleon was coming up after him. Blücher withdrew to Oulchy-le-Château, on the northern side of the river, breaking the Marne bridges behind him. This was a vital step, as the Emperor had no bridging train.

Although he had no firm knowledge of the exact locations of Bülow and Winzingerode, Blücher had a general idea and headed northwards for Laon, where Bülow actually was at that point. Bülow advanced to the south-west, to Soissons on the River Aisne and took command of the vital bridge there when the timid General Moreau capitulated with the garrison and was allowed to march off to fight another day. Ignorant of this development, Blücher, with Napoleon in hot pursuit, crossed to the northern bank of the flooded Aisne east of Soissons and joined up with Bülow and Winzingerode on 4 and 5 March. He now had about 100,000 men under his command. Napoleon's great plan had been foiled. To make matters even worse, the Emperor received news that Macdonald was ill and that Troyes was in enemy hands again. 'I cannot believe such ineptitude,' he raged. 'No man can be worse seconded than I am!'

Racing after his arch enemy, Napoleon clashed with Voronzov and Sacken at Craonne, 15 miles south-east of Laon, on 7 March. Ney's assault was premature and failed. Napoleon attacked again at Laon on 9 March, but achieved nothing. That night, having learned that Marmont's VI Corps of 10,000 was isolated east of Napoleon's main body, Blücher attacked him and inflicted a major defeat on him for the loss of about 750 men. Luckily for Napoleon, Blücher then suddenly fell ill again and the vibrant energy that his fiery presence breathed into his commanders and into his men vanished. No concerted follow up took place and Napoleon withdrew relatively intact.

Foiled yet again, the Emperor learned that St. Priest's corps was isolated at Reims to the east. He swept down on 13 March and threw him out of the town with loss. At this point

The Battle of Fère-Champenoise

Schwarzenberg, with 90,000 men, was advancing through Nogent down the Seine on Paris, opposed only by Macdonald and 30,000. Leaving Marmont and Mortier with 21,500 men to watch Blücher, Napoleon took 30,000 and rushed south-west to strike the Army of Bohemia in flank. On 20/21 March he attacked Schwarzenberg at Arcis-sur-Aube, but was utterly surprised at the firmness shown by his adversary and was repulsed. The losses the French suffered are unknown. The allies lost in the region of 3,000 men.

The French fell back to the north-east and Napoleon stopped in St. Dizier. By 25 March Blücher and Schwarzenberg were in contact along the Marne and between Napoleon and his capital. Recently captured despatches revealed to the allies the panic-stricken, unprepared state of Paris and the dispositions of the remaining French troops in the area. At Czar Alexander's insistence, they decided to strike at Marmont and Mortier, who stood at Fère-Champenoise, between them and the enemy capital, and to ignore whatever Napoleon might try to do to their lines of communication. The czar had kept a clear head and had divined the critical move that would, at last, end the wavering campaign.

Unaware of this leak of critically important intelligence, Napoleon remained at St. Dizier from 24–28 March, a fatal delay. He confidently expected the panic-stricken allies to rush eastwards to escape his usual *manoeuvre sur les derrières*, as had so often been the case in the past. But this time there was little alarm in the allied camp; they knew the real score. Sending Winzingerode's corps of 8,000 cavalry and 46 guns to mask Napoleon, Alexander convinced the allied high command to direct the Armies of Bohemia and Silesia to march on Paris. This time it was Napoleon's turn to dance to the allied tune

Fère-Champenoise lies south-east of Étoges, just east of the source of the River Grand Morin, and 70 miles east of Paris. It was nine miles to the west of Sommesous, which area was occupied by Marmont's VI Corps and Mortier's Young Guard with 16,700 men and 84 guns. On 23 March Mortier, in Château-Thierry, had received the Emperor's order to join with Marmont as quickly as possible and now they were both hurrying eastwards. On 25 March they met the allies at the villages of Soude-St.-Croix and Soude-Notre-Dame on the road from Troyes to Châlons; the Soude stream covered their front. The point cavalry unit of the allied advanced guard, under Prince Adam of Württemberg, surprised a French outpost and took prisoners, who confirmed the presence of the two marshals. Prince Adam brought up two guns and fired a few shots to see what he could flush out. In response the French rapidly formed line of battle.

Between about 8 and 9 o'clock, Crown Prince Wilhelm of Württemberg came on the scene with the rest of the cavalry of the advanced guard. Without awaiting the arrival of the infantry, he ordered a cavalry attack. Count Piotr Pahlen, with his Cossacks and Russian hussars plus Kretov's cuirassiers, was to take the right, the Crown Prince led the left. The intention was to threaten both enemy flanks rather than to risk a frontal assault. After only a brief volley, the French line fell back on Sommesous, though in the process the allied cavalry caught the rearguard in the open and cut many of them down.

The French formed line again at the village. Mortier's Guard, slightly advanced, was left of Sommesous, Marmont's corps to the right. Crown Prince Wilhelm brought up his artillery to soften up the opposing line. A sharp fire-fight developed, during which Mortier fell back into line with his ally, on a slight ridge behind the Somme stream and between the villages of Vassimont and Montepreux.

By now Count Nostitz's Austrian Kürassier division had come up. Count Pahlen, with Kretov's division, sought to find a way through the swampy ground which covered the French right. Prince Adam's allied brigade made ready to deliver a frontal assault. In the first line were Nostitz's division, formed in column of divisions (that is with a frontage of two squadrons); Prince Adam's troops deployed to either flank. It was midday.

The first allied regiments to hit the French were the Austrian Erzherzog Ferdinand Hussars and the Württemberg Prinz Adam Jägers. Two squadrons of the latter had been deployed in a skirmishing cloud ahead of the main line of battle. They clashed with skirmishers of the French hussars and chasseurs and overthrew them after a brief fight. The French cavalry fell back, reformed and tried to come on again, but were repeatedly

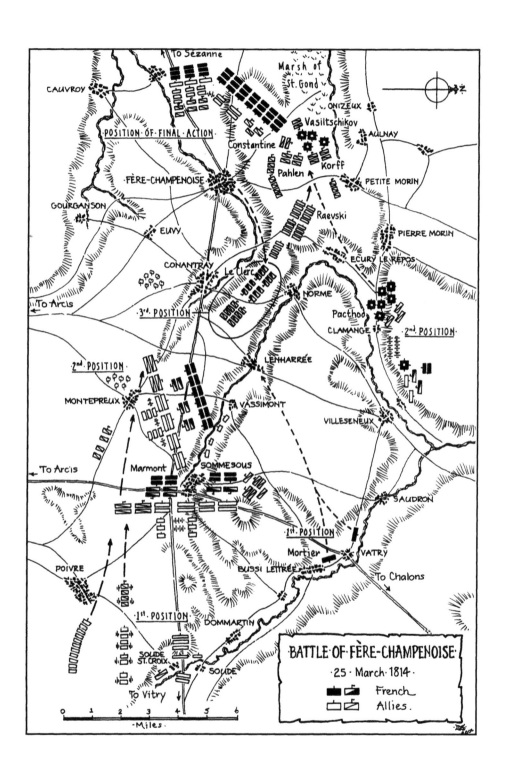

pushed back until, losing heart, they withdrew and were apparently replaced by General Belliard's 'Spanish' dragoons (5th, 6th, 21st, 25th and 26th Dragoons and 23rd Chasseurs). These advanced and extended their skirmishing line so as to envelop the left flank of the Württembergers. General von Jett countered this by deploying his two remaining squadrons of the Prinz Adam Jägers. The skirmishing continued at a hot pace, in which the Spanish dragoons were repeatedly bested and pushed back.

Meanwhile, in the centre of the field, General von Nostitz led a charge into the enemy batteries. Delfours's brigade was hit by a salvo of canister at close range and thrown into confusion. In attempting to fall back, their confusion was turned into chaos by their unhandy column of divisions, and they were charged by Bordesoulle's cuirassiers and swept away. The Liechtenstein Kürassiers were now surrounded by enemy cavalry and caught in a desperate mêlée. This reverse in the centre also caused the Prinz Adam regiment to retire.

The Crown Prince set himself at the head of the Erherzog Ferdinand Hussars and led a counter charge. The Prinz Adam Jägers returned to the fight and the 5th Jägers-zu-Pferde joined in the charge. This restored the front and freed the Liechtenstein Kürassiers from their predicament. Now the regiments on both sides had to pause and re-order themselves, a process which took some time.

While this was going on, Mortier and Marmont withdrew to a new line between Conantray and Clamange; their infantry in squares in chequerboard formation, the cavalry in the rear. Meanwhile, the Crown Prince received a despatch that Grand Duke Constantine himself was coming up from Sommepuis via Poivre to take command of the cavalry of the Russian Imperial Guard to take the enemy in the right flank. Pahlen joined Prince Wilhelm and Nostitz.

The united allied cavalry moved forward against the enemy. The Prinz Adam Jägers charged a square formed of about 1,000 Young Guard Tirailleurs but the steady infantrymen drove them off. A second charge also failed to shake the French guardsmen. Yet a third charge was made and this time the square was momentarily broken but reformed, although two guns at the side

The Battle of Fère-Champenoise

of it were taken by the Württembergers. They were now joined by the Ferdinand Hussars and a fourth charge was made, led by General von Jett.

A violent rain and hail storm burst over the field, driving into the faces of the French just before the cavalry hit them. This time the square broke. The tirailleurs were scattered and most of them cut down. The French began to fall back through Conantray. As they did so, the Austrian Ferdinand Hussars and Kaiser Kürassiers and the Cuirassiers of the Russian Imperial Guard charged and broke a square of the Voltigeurs of the Guard. The village was utterly blocked with abandoned vehicles and other debris of the combat. The allied cavalry had to skirt around it, having to jump the stream one by one wherever they could. The sodden ground meant that the Württemberg artillery teams, which had been on the move and in action for ten hours, were now too exhausted to move any further.

It was now about 2 o'clock and the French troops had finally had enough. They streamed away westwards around Fère-Champenoise and on to the heights of Broussy-le-Grande, St. Loup and Linthes in the direction of Sézanne. Having reorganised themselves after passing the defile of Conantray, Pahlen and Prince Adam prepared to advance to take the French in their left flank, while the Grand Duke Constantine moved to hit their right. There were now almost 10,000 allied cavalry on the field.

At this point, cannon fire was heard to the rear of the allied forces. It was Generals Amey and Pacthod, coming south from Étoges to join the two marshals with their Garde Nationale divisions. They had run into the forces of the Army of Silesia and were being destroyed but Marmont and Mortier assumed that the gunfire meant that Napoleon was advancing to save them. They turned back towards the east. Their cavalry poured over a ridge to find three Russian guns in their path and were quickly on them with shouts of 'Vive l'Empereur!' Seeing this disaster, Oberstleutnant von Reinhard at once threw his tired Prinz Adam Jägers against them and an Austrian Kürassier regiment joined in. The French cavalry hesitated, then fell back. The guns were saved. Darkness put an end to the action. The allies were too tired to push on.

The Battle of Fère-Champenoise

Fère-Champenoise is a rare instance of an action in which a force of about 10,000 cavalry, with some artillery support, but no infantry (they were in rear, but never caught up with the action) soundly trounced an opponent twice their number, having plentiful infantry, much of it from the élite Imperial Guard, commanded by two very successful and experienced marshals of France. The piece is piquant in that the Württemberg cavalry had fought with the French as faithful allies from 1805 until the end of 1813. That they should now perform so effectively against their old comrades proves the depth and reality of the sense of German nationalism that had been awakened.

Marmont and Mortier withdrew to Sézanne. They had lost 5,000 killed and wounded, and, with those taken with Amey and Pacthod, 10,000 prisoners, 80 guns and 250 ammunition and baggage wagons. Most significantly, the allies had broken two squares of Napoleon's much-vaunted Imperial Guard. Of these spoils, the allied cavalry of the IV (Württemberg) and VI Corps took 15 guns during the action. A further 30 abandoned pieces, together with 100 ammunition and baggage wagons and 4,000 prisoners, were also taken by them after the clash. The allied cavalry had lost about 2,000 killed, wounded and missing. The marshals had suffered the loss of about half their troops and almost their entire artillery and baggage train. They were in no shape to fight again for some time, and time was the one thing that they did not have.

Next day the allied advance on Paris continued, against only faint resistance. Meanwhile, the scales had fallen from Napoleon's eyes. He had aimed to lead the allies off to the east and to pick up the garrisons of several border fortresses to swell his ranks. His marshals were more realistic and refused to obey. Early on 28 March Napoleon at last saw reason and began his race to save Paris. He was too late. The city fell on 30 March when he was still at Troyes. Talleyrand began negotiations with the allied monarchs and the marshals had all had enough. They could see the true hopelessness of their situation. Only some of the lower ranks clamoured for Napoleon to lead them on to yet another crushing victory. He believed that their cheers and chants could change his fate and called his marshals to receive their orders for the next master-stroke.

The Battle of Fère-Champenoise

'The army will not march,' answered Marshal Ney. 'The army will obey me,' replied Napoleon. 'The army will obey its commanders,' retorted Ney defiantly. Ney was right. On 4 April Napoleon offered to abdicate in favour of his son, the King of Rome. The allies rejected the Emperor's offer and demanded his unconditional abdication; they had upped the stakes. Next day, Marmont took the shattered remnants of his corps over to the allies. On 6 April Napoleon abdicated unconditionally. He had no cards left to play. The campaign of 1814 was over but the Napoleonic wars were not.

Sources

Alison, *A History of Europe*, London, 1848

Bodart, Gaston, *Militär-historisches Kriegs-lexikon*, Vienna and Leipzig, 1908

Chandler, David, *The Campaigns of Napoleon*, London, 1966

Esposito, V.J., and Elting, J.R., *An Atlas of the Napoleonic Wars*, London, 1999

Martinien, A., *Tableaux pars Corps et pars Batailles des Officiers Tués et Blessés Pendant les Guerres de l'Empire 1805–1815*, Paris, 1890

Six, Georges, *Dictionnaire Biographique des Generaux & Amiraux Français de la Revolution et de l'Empire (1792–1814)*, Paris, 1934

Starklof, R, *Geschichte des Königlich Württembergischen vierten Reiterregiments Königin Olga 1805–1866*, Stuttgart, 1867

CHAPTER 14
The Battle of Waterloo
18 JUNE 1815

The strategic scenario of this campaign was very familiar. Napoleon, holding absolute political and military power in France, aimed to exploit the advantages of internal lines of communication, with a homogeneous army, against the usual disparate array of mutually mistrustful allies, operating on external lines of communication, speaking three or more different languages and using two different calendars. Additionally, as usual, the allied military commanders were each subject to political masters who might interfere negatively with sound military plans for their own reasons.

The orders of battle for this epic battle are so vast that we have no room for them here. Details given in the appendices (see pages 298–99) will be limited to d'Erlon's I Corps, the French cavalry and the British Heavy and Union Cavalry Brigades.

Louis XVIII, 'Louis the Unavoidable' as he was known, had been placed on the throne of France by the allies in 1814. He rapidly became widely unpopular with many Frenchmen and demoralised the army by carefully destroying the traditions which had become sacred to them during the Glory Years. Unemployed soldiers and thousands of officers struggling to get by on half pay formed a reservoir of discontented warriors who had little to lose in post-war France. They yearned to be swaggering once more through the streets of conquered foreign capitals, where all was theirs for the taking. The peasantry feared the re-introduction of privileges for the returned émigrés and the church, that their land would be confiscated and that the old feudal duties, abolished in 1792, would be re-imposed. All this was bad enough but perhaps Louis's greatest error had been to fail to pay Napoleon his agreed pension of two million francs.

Napoleon, restless on the tiny island of Elba, bored, cut off from his wife and son (Metternich confiscated all letters that Marie-Louise wrote to him), running out of cash, and well informed of Louis's mounting unpopularity, decided to risk everything on a last gamble. Not for him an end with a whimper; he was going to go out with one almighty bang. On 26 February 1815 Napoleon left Elba, accompanied by Generals Bertrand, Cambronne and Drouot and a thousand men of the 'Elba Guard'. On 1 March he landed in France and began a rapid march on Paris.

We will pass over his legendary triumphal march to the capital, the collapse of the king's power base and his flight to Belgium on 19 March. Napoleon entered the Tuilleries next day and took over the reins of state. He was back. But the old magic of yesteryear now worked less strongly and in a more restricted circle than before. Several of his marshals and generals had had enough of him and had accompanied Louis into exile; the Chamber of Deputies remained cool. With his customary energy and genius Napoleon set to work at once on the mighty tasks that confronted him. The allies were in conference in Vienna, squabbling over the redistribution of Europe.

Joachim Murat, the Emperor's brother-in-law and King of Naples, had attacked the Austrians at Modena in northern Italy on 4 April in support of Napoleon, a support that was uncoordinated with the Emperor and unsought for, following Murat's treachery in 1814. By mid-May Murat had been convincingly defeated and was a fugitive.

By 1 June 1815 the allies were in the process of forming the following armies along France's borders: in the north, around Brussels, was the Duke of Wellington's Anglo-Dutch army (including a large German contingent) of 108,000 men, of whom 94,600 were available for use in the field; to the east of them around Liège and Namur, was Blücher with 123,000 Prussians; on the River Moselle was General Kleist von Nollendorf's IV Prussian Corps of 26,200; along the upper Rhine, from

Mannheim to Basle, was Field Marshal Prince Schwarzenberg with 225,000 Austrians and Germans; in Piedmont was General der Kavallerie Frimont with 60,000 Austrian and Italian troops; Feldmarschall-Leutnant Bianchi was in central Italy with 25,000 more Austrian and Neapolitan troops; and finally marching west from Poland was Field Marshal Barclay de Tolly with 168,000 men. Not all of these armies were in equally advanced states of readiness. Those of Wellington and Blücher led the pack, while the Russians and the troops in Italy brought up the rear.

To counter these threats, Napoleon launched a diplomatic campaign designed to disarm and to divide his enemies. They had heard it all so many times before. Although there were those who began to be beguiled by his protestations of being a good boy in future, others, led by Britain and Prussia, were determined to put an end to his disruptive career once and for all.

The greatest threat to Napoleon lay to the north, in Belgium, and he knew it. As usual, the Emperor's intelligence network was working at top speed to fill in the details of the activities and intentions of his enemies. In addition, he was actively clamping down on all possible leaks of information from France, which might tell the allies what he intended to do. In the weeks prior to his attack on Blücher and Wellington, France's borders were sealed, the press muzzled and postal services curtailed. Napoleon effectively placed France in purdah.

To counter the allied threats, Napoleon formed the following armies: Armée du Nord, along the Belgian border, 128,000 men under his own direct command; Armée du Rhin, under General Count Rapp, facing Schwarzenberg, 23,100 men; Corps d'Observation du Jura, under General Count Lecourbe, facing Switzerland, 8,400 men; Armée des Alpes, 23,600 men under Marshal Suchet, opposite Piedmont; Corps d'Observation sur le Var, under Marshal Brune, 5,500 men on the Riviera; Corps d'Observation des Pyrénées Orientales, 7,600 men under General Count Decaen; Corps d'Observation des Pyrénées Occidentales, 6,800 men under General Count Clauzel; Armée de la Loire, 11,000 men in the Vendée under General Baron Lamarque, already fighting to put down a royalist uprising there. The Channel coast was left to the care of the coastguards. Britain's only available field army was already in Belgium; there would be

The Battle of Waterloo

no other invasion. Given the distribution of these forces, it was obvious that Napoleon's plan was to act offensively against the allies in the Netherlands and defensively on all other fronts.

Having made this very simple decision, Napoleon moved with all his customary speed to close with and destroy his enemy. Wellington's lines of communication ran west to the coast, those of the Prince of Orange (appointed as Wellington's deputy) ran north to Brussels, Blücher's went off to the east. One great surprise blow at the junction of this brittle collection of forces must burst them apart and his problems would be solved. With Britain and Prussia defeated and humiliated, the alliance against him would be brought to heel. He would be back in the saddle again. His concentration of his army up against the Belgian border was a masterpiece of stealth and security. The allies knew that he was up to something, but his exact dispositions and aims were unclear. On 15 June he surprised Zieten's I Prussian Corps at Fleurus, Gilly and Gosselies while it was still extended. Zieten fell back to Ligny to join up with Blücher's main body which was concentrating there.

Next day saw the twin battles of Ligny and Quatre Bras. At the former, Blücher's Prussians were bloodily defeated by Napoleon and the old warrior was himself injured after being trapped under his horse for a time. Despite the serious situation in which the Prussian Army now was, Blücher's chief of staff, General von Gneisenau, commanding for the injured Blücher, overcame his mistrust of the British, and ordered the battered army to withdraw north on Wavre, to stay in contact with his ally, instead of heading for home and safety towards the east. There is no doubt that this courageous decision saved the coalition.

At Quatre-Bras that same day, the Dutch, British and Brunswickers had held a vital crossroads – just. They were much assisted by Marshal Ney, the opposing commander. Ney had joined Napoleon only the day before and was unfamiliar with his formations, commanders and staff. His actions were uncharacteristically slow and timid. A tug of war developed between him and Napoleon over the employment of d'Erlon's I Corps, with the result that it fought at neither battlefield that day.

Wellington, too, had been surprised by Napoleon's excellent

security precautions. Now he recovered and scrambled back to occupy the ridge at Mont St. Jean, some 11 miles south of Brussels. He had previously picked this site out as a potential blocking position and was to exploit its advantages to the full in his familiar Peninsula style. As usual, he would fight a defensive battle, using all the possibilities offered by the features of the ground, the hedges, trees and farms upon it. He drew up most of his line behind the crest of the ridge. Bijlandt's Netherlands brigade was, it seems, forgotten, and was on the scarp of the ridge, just to the east of the fortified farm of La Haye Sainte only about 1,000 yards from the great battery which Napoleon would set up. Saxe-Weimar's brigade was also on the face of the ridge, but was off to the east, behind the villages of Papelotte and Frichermont, which were held by allied troops. In front of the right wing, Wellington had fortified the château of Hougomont. This was to be a crucial bulwark against the tide of the French offensive. To protect his line of communication, Wellington had placed 17,000 men in Hal, some 10 miles off to the west.

Leaving Marshal Grouchy with 33,000 men to follow up the defeated Prussians and keep them away from the British, Napoleon led about 77,000 men of the Imperial Guard, I, II, VI Corps and the III and IV Cavalry Corps to smash his way through Wellington's polyglot 68,000 at Waterloo. Luckily for Wellington, it had rained hard and long on 17 June. Napoleon was unable to deploy his artillery as he wished, and misspent precious time wallowing in choruses of 'Vive l'Empereur!' as he toured his adoring troops accompanied by massed bands. 'Ask of me anything but time.'

As the clock ticked, the battered Prussians (89,000 men) were marching west from Wavre, to take the French in flank. Grouchy was fumbling to find them. It was far too late when he at last made contact in Wavre at about 7 o'clock that night, when he bumped into General Thielmann's corps.

Napoleon deployed his forces from west to east as follows. Reille's II Corps (23,000 men and 46 guns) to mask Hougomont and to extend to the Brussels road; right of the road was d'Erlon's I Corps (19,000 men and 46 guns); Lobau's VI Corps (9,084 men and 38 guns) and the Imperial Guard under Mortier (19,428 men and 96 guns) were behind the centre, ready to push up the main road to Brussels. Kellermann's III Cavalry Corps (3,555 men and 12 guns) was behind the left flank, Milhaud's IV Cavalry Corps (2,869 men and 12 guns) was behind the right flank. Reille was ordered to occupy the wood south of Hougomont to distract the allies but he did more than he was ordered and became involved in a desperate struggle for possession of the place, starting at 11.20.

The French bombardment opened at 12 o'clock from a great array of three 12-pounder batteries and four 6-pounder batteries – 54 guns in all – that had at last been dragged into position to the east of the Brussels road, facing La Haye Sainte. At the same moment, Prussian troops were approaching Chapelle St. Lambert, only some four miles off to the east of the Charleroi–Brussels road. Time was running out for the Emperor very quickly indeed. Napoleon weighed his options. There were only two: fight or run. With the strategic odds against him, only a crushing victory – here and now – would do. To withdraw would merely act against him, as the coalition forces grew stronger by the day. He was also dismissive of Wellington's skills as a commander and ignored the warnings given to him by those of his generals who had been defeated by Wellington. Fight it would be, the compulsive gambler's last throw, everything on one card.

The Charge of the Guards and Union Brigades

At about 1.45 in the afternoon d'Erlon's I Corps, and Milhaud's IV Cavalry Corps were launched by Ney, in two separate assaults, up the Brussels road against La Haye Sainte. Neither was supported by units of artillery, cavalry or infantry respectively. Bijlandt's unfortunates, who had endured losses at Quatre-Bras and in the great pounding of the French artillery since that noontime, broke and fled to the rear as the advance rolled in. It is difficult to blame them. Wellington's skirmishers were pushed back over the crest of the ridge.

It was at this point that General Lord Edward Somerset's Guards Cavalry Brigade and General Sir William Ponsonby's Union Brigade were launched to break up the assault of d'Erlon's corps. Somerset's regiments were the 1st and 2nd Life Guards, the Royal Horse Guards (Blues) and the 1st (King's) Dragoon

Guards. Somerset's men clashed with the cuirassiers of Travers's brigade of Wathier's 13th Cavalry Division of Milhaud's corps (7th and 12th Cuirassiers), veterans whose battle honours included actions such as Ulm, Jena, Eylau, Friedland, Aspern-Essling, Wagram, Borodino and Dresden. They advanced with the sure air of men expecting to win again, particularly as they had seen Bijlandt's brigade disappear over the ridge. As they crested the ridge, the cuirassiers were met with fire from the batteries of Ross and Lloyd. They staggered, collected themselves and came on. Their trumpeters sounded the charge. With the usual shouts of 'Vive l'Empereur!' the steel-clad mass rolled forward to crush their opponents. As they approached the first line of British infantry, they were met with volleys of musketry, which brought down numerous men and horses. It did not slow them; they trotted on.

Somerset's regiments were at a gallop by this time. The foes crashed into one another with incredible force, both fully convinced of their own superiority. The swords of the cuirassiers were longer than those of the British, and they had the advantage of body armour. To negate these advantages, the British sought to close with their opponents as fast as possible. Man hacked and stabbed at man, horses reared, riders fell into the storm and were trampled underfoot. It was bloody chaos for a few moments. Then suddenly the French broke, overthrown by the power of the British charge.

On the French right, meanwhile, their charge had been brought to a halt by a sunken road, which they had to cross. As they swirled in confusion down into the obstacle and up the opposite bank, they saw, with horror, the 2nd Life Guards thundering at them, swords at the point. To stand would be to die. The cuirassiers scrambled back down into the road and fled off to their right, across the Charleroi road and across the front of the 95th Foot, hotly pursued by the 2nd Life Guards. When they reached the line of French infantry skirmishers, they rallied and took up the fight with their pursuers, but were overwhelmed and either fled or were taken. Corporal Shaw of the 2nd Life Guards was a noted boxer of great strength. In the space of a few minutes, he cut down no fewer than nine cuirassiers, before being shot and killed in his turn. The 1st Life Guards in the sunken road now came up past Bachelu's 5th Division of Reille's II Corps, to the west of the Brussels road. Here, the British cavalry were caught in a destructive fire to which they had no effective answer. Taking casualties, they fled back up the road to the ridge.

When Somerset led his Guards Brigade off, Ponsonby followed to their left with his Union Brigade, in accordance with his orders. However, not being certain of the terrain and the tactical situation to his front, he halted his men before the Wavre road and rode forward to reconnoitre the situation. To the eastern side of the Brussels road, the rear brigade of Allix's 1st Division of d'Erlon's corps was advancing up the slope in two columns, each of two battalions. These were the 54th and 55th Line. They were in support of the other brigade, the 28th and 105th. Similarly, the leading brigade of Marcognet's 3rd Division (25th and 45th) was being supported by the 21st and 46th Line, also in two columns. At this point in the battle, Kempt's 8th Brigade of Picton's 5th Division was advancing down the ridge to the west of Allix's and Marcognet's columns, which had just crested the ridge. Allix's men found no infantry left before them but Marcognet's men received very destructive fire from Captain von Rettberg's Hanoverian battery as they advanced. When they came over the ridge, they were confronted by the remnants, some 230 men, of the 92nd Highlanders, a regiment which had suffered badly at Quatre-Bras. The Highlanders were facing about 2,000 Frenchmen.

Major-General Sir Denis Pack, commander of the 9th Brigade (1st and 44th Foot, 42nd and 92nd Highlanders) was with the Highlanders, and saw the mass of the enemy breaking through the hedge in front of him. He made a very quick decision. Turning to his men, he said: '92nd, you must charge – all in front of you have given way!' With a loud cheer, the 92nd went forward with a will, pipes playing. They were met with a volley by the head of the column. They did not reply, but marched on. At about 20 paces distance, a sudden panic seized the leading French infantry and they scrambled back through the hedge, throwing the entire column into hopeless confusion. The 92nd stopped and poured in a destructive volley and charged.

At this point, Ponsonby's cavalry came up. As the Scots Greys passed by the 42nd and 92nd, many of the infantry grabbed hold of their countrymen's stirrup leathers and hitched a lift so as to be at the enemy all the faster. Marcognet's men were surprised and shocked, frozen at the sudden defeat of their column head and the appearance of cavalry. They had no chance and were swept away. A small group of the 45th stayed, clustered about their sacred eagle, with the battle honours Austerlitz, Jena, Friedland, Essling and Wagram on it. Sergeant Ewart of the Scots Greys saw the trophy and made for it. After a desperate struggle, he cut down the Eagle Guard and made off with his prize. He was directed to take it back to Brussels where he received a hero's welcome. The eagle was later displayed in the Royal Hospital, Chelsea; Ewart received a commission in the 3rd Royal Veteran Battalion in 1816.

Meanwhile, the Scots Greys careered on to assault Marcognet's other brigade. Their morale had already been broken. Some of them managed to get off a ragged volley, then they went down under the heavy swords of the Scots. Hundreds were cut down; hundreds captured. On the right, the Royal Dragoons had been equally successful in charging the head of Allix's column. These managed to get off a scattered volley, which brought down about 20 troopers and then the cavalry struck them in front and flank and crushed them into a panic-stricken mob, running for their lives.

In the chaos, Captain Clark of the Royal Dragoons espied the regimental eagle of the 105th Line. It bore the battle honours of Jena, Eylau, Eggmuhl, Essling and Wagram. He shouted out: 'Right shoulders forward – attack the colour!' Hacking his way through the mob, Clark ran the *port-aigle* through and grabbed at the colour as it fell. Only able to grasp a fringe of the cloth, he would probably have dropped it, but Corporal Stiles of his regiment came up and secured it. Clark then tried to break the gilt eagle from the staff. 'Pray, Sir, do not break it,' said Stiles. 'Very well,' said Clark, 'Take it to the rear as fast as you can – it belongs to me.' This eagle was deposited in Royal Hospital Chelsea; Clark was appointed a Companion of the Order of the Bath and Stiles awarded a commission in the 6th West India Regiment.

The third regiment of the Union Brigade was the 6th

The Battle of Waterloo

Dragoons (Inniskillings). As reserve of the brigade, it was to the rear of the other two initially. Moving forward, the Inniskillings burst through the hedge along the ridge and charged down on Allix's support brigade, the 54th and 55th Line. With a loud 'Hurrah!' the Irishmen poured 100 yards down the slope. The right and centre squadrons hit the 55th; the left squadron tackled the 54th. The startled French were only able to get off an incomplete volley before the cavalry were among them, hacking and stabbing to right and left. The terrified infantrymen scattered and ran. Hundreds were captured, hundreds more killed or wounded. The I French Corps lost over 3,000 men in this interlude. It would be 4 o'clock before it would be fit to be used again. 'Ground I may recover; time, never.'

At this point, the 2nd Life Guards, the 1st Dragoon Guards and the Union Brigade, drunk with success and out of control, careered up into the French position, into the gun lines, cutting down any gunners they could reach. Only the Blues were collected and under control, but they were far to the rear of the steeplechase. Lord Uxbridge, overall commander of Wellington's cavalry, had the halt and rally sounded repeatedly. None heeded the signals. Instead they charged onward. Somerset and Ponsonby had lost control of their officers and men. The King's Dragoon Guards came under fire from Bachelu's men and from artillery on their right. The French cuirassiers began moving in on the British cavalry. The British were receiving more and more artillery fire, and now they saw Jacquinot's lancer brigade of d'Erlon's corps (who had not been sent in to support their infantry comrades' disastrous attack) bearing down on them.

At last, the peril of their exposed situation dawned on the British. Discretion – too late – began to take the place of valour. They turned back for their own lines, but their horses were now blown and exhausted and could scarcely manage a trot. Somerset's brigade was lucky and managed to get back to its own lines without much damage. The Union Brigade, and the Scots Greys in particular, were less fortunate. They were now tired out and in complete confusion, regiments, squadrons and troops all mixed up together. Farine's cuirassier brigade (1st Brigade 14th Division) charged them from the front, whilst Jacquinot's lancers hit them in the flank; it was a perfect trap. The vengeful lancers had a field day. At last, Vandeleur's 4th Cavalry Brigade (11th, 12th and 16th Light Dragoons) came to the rescue on General Ponsonby's left flank. The leading regiment was the 12th, commanded by Lieutenant-Colonel Frederick Ponsonby. It struck Marcognet's last intact column and ruined it. The 12th and 16th Light Dragoons then attacked the lancers to save the survivors of the Union Brigade.

For Sir William Ponsonby, the day ended badly; he was killed by one of the lancers. His namesake fared little better. He was wounded seven times and left for dead on the field.

This dramatic cavalry action demonstrated all the frightening power of mounted troops against infantry but it also showed how cavalry ought *not* to be used.

The Grand Failure

The great battery was now reinforced and began to pound Wellington's centre again. Casualties mounted and men began to trickle off to the rear. Napoleon was punching his way through to Brussels.

At about 4 o'clock, Ney, who had been studying the weakening of the allied centre, ordered Milhaud's IV Cavalry Corps to charge into it, just to the west of the Brussels road. Somehow, though it is not quite clear how, Lefebvre-Desnouëttes's Light Cavalry Division of the Imperial Guard joined the charge. As luck would have it, they struck a part of Wellington's line which had not been badly damaged by the French artillery. Seeing the great storm approaching, Wellington formed his infantry into a double line of squares, in a chequerboard pattern, on the reverse slope of the ridge. As tactics dictated, the great battery had to limit its fire as their own cavalry swarmed up the slope of the enemy position. The approach of the French cavalry forced the British gunners to abandon their guns and withdraw into the safety of the squares after they had fired a last salvo into the enemy cavalry.

Over 3,600 cavalry trotted across the muddy valley and up the slope, closing to the centre as the allied guns tore gaps in their ranks. They were through the allied gun lines and ready to break the infantry line, when they were confronted by the squares of steady infantry, each ready to receive them with deadly volleys. The Earl of Uxbridge led the allied cavalry in a counter-charge which threw the French back over the ridge. They came on again and forced Uxbridge back in his turn.

Napoleon had by now heard that the Prussians were advancing on Plancenoit in force and that Grouchy was too far off to affect the battle at all. It was time for the last throw, good money after bad. He ordered Kellermann's III Cavalry Corps and Guyot's Heavy Cavalry Brigade of the Imperial Guard to join the hopeless mêlée. By throwing even more cavalry at the unbroken centre of a hostile line, unsupported by artillery or infantry, Napoleon was condemning many of them to a pointless death. Worse than that, he was wasting his own precious assets.

Many of Napoleon's apologists attribute his mediocre performance on 18 June to his inflamed bladder, his piles, other mysterious ailments, his incompetent subordinates, his chief of staff, the weather or any one of a dozen other factors. Few take into account that Melas, at age 70 in 1800, put up a better performance than the Emperor at age 46 at Waterloo. Perhaps his mortality was beginning to show.

So, Napoleon ordered the cavalry of his Imperial Guard into a pointless, useless, suicidal charge – or did he? A quotation from a letter written in 1835 by ex-captain Fortune Brack, of the 2nd Chevaux-légers Lanciers of the Guard, a participant in the battle, reveals a new aspect of this futile exercise.

> 'Impassioned by our recent success against Ponsonby, and by the forward movement that I had noticed being executed by the cuirassiers to our right, I exclaimed, "The English are lost! The position on which they have been thrown back makes it clear. They can only retreat by one narrow road confined between impassable woods. One broken stone on this road and their entire army will be ours! Either their general is the most ignorant of officers, or he has lost his head! The English will realise their situation – there – look – they have uncoupled their guns."
>
> I was ignorant of the fact that the English batteries usually fought uncoupled.
>
> I spoke loudly, and my words were overheard. From the front of our regiment a few officers pushed forward to join our group. The right hand file of our regimental line followed

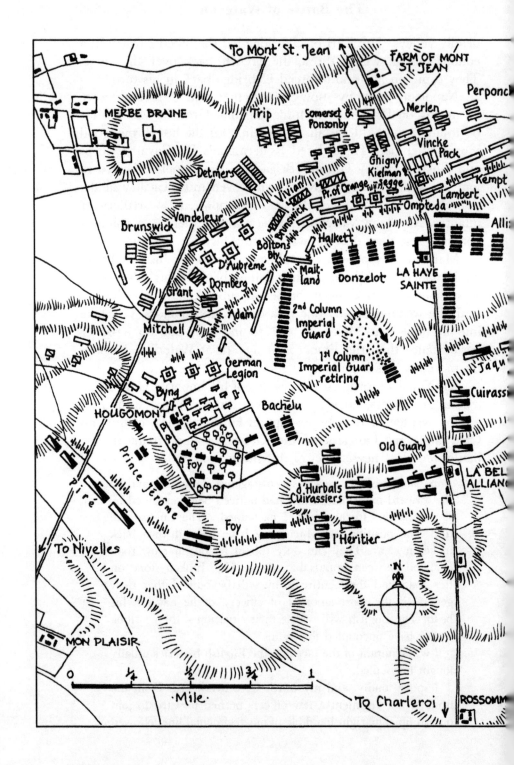

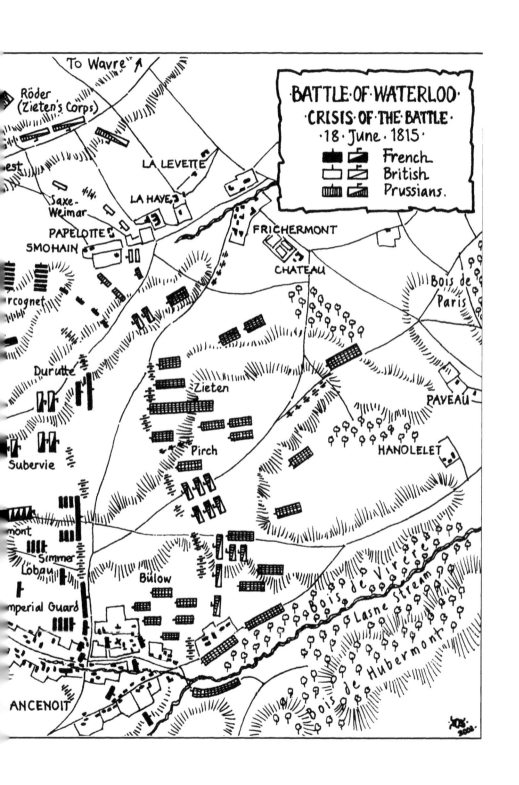

them; the movement was copied in the squadrons to the left to restore the alignment, and then by the Chasseurs-à-Cheval of the Guard. This movement, of only a few paces at the right, became more marked [as it passed] to the left. The brigade of the Dragoons and Grenadiers-à-Cheval, who were awaiting the order to charge at any moment, believed this had been given.

They set off – and we followed!

That is how the charge of the Imperial Guard cavalry took place, over the reason for which so many writers have argued so variously.

From that moment, lining up to the left, we crossed the road diagonally so as to have the whole Guard cavalry on the left side of this road. We crossed the flat ground, climbed up the slope of the plateau upon which the English army was drawn up, and attacked together.

The order in which that army was drawn up, or the part exposed to our view, was as follows:

To its right were the Scots Foot [Guards?], close to undergrowth which extended to the bottom of the slope. This infantry delivered heavy and well-directed fire.

Then came the squares of line infantry, ordered in a chequerboard pattern, then similar squares of Hanoverian Light infantry; then a fortified farm [La Haye Sainte].

Between the squares were uncoupled batteries, whose gunners were firing and then hiding under their guns, behind them some infantry and some cavalry.

We were nearly level with this farm, between which and us our cuirassiers were charging. We rode through the batteries, which we were unable to drag back with us.

We turned back and threatened the squares, which put up a most honourable resistance.

Some of them had such coolness, that they were still firing ordered volleys by rank.

It has been said that the Dragoons and Grenadiers-à-Cheval to our left broke several squares; personally I did not see it – and I can state that we Lancers did not have the same luck, and that we crossed our lances with the English bayonets in vain. Many of our troopers threw their weapons like spears into the front ranks to try to open up the squares.

The expenditure of ammunition by the English front line

The Battle of Waterloo

and the compact pattern of the squares which composed it meant that the firing was at point-blank range, but it was the harm which the artillery and the squares in the second line were doing to us, in the absence of infantry and artillery to support our attack, which determined our retreat.

We moved slowly and faced front again in our position at the bottom of the slope, so that we could just make out the first English line.

It was then that Marshal Ney, alone and without a single member of his staff accompanying him, rode along our front and harangued us, calling out to the officers he knew by their names. His face was distracted, and he cried out again and again: "Frenchmen, let us stand firm! It is here that the keys to our freedom are lying!" I quote him word for word.

Five times we repeated the charge; but since the conditions remained unchanged, we returned to our position at the rear five times.

There, at 150 paces from the enemy infantry, we were exposed to to the most murderous fire. Our men began to lose heart. They were being hit at the same time by bullets from the front and by cannonballs from the flank, and by new projectiles (small shells) which exploded over their heads and fell. These were shrapnel shells and we had not come against them before.

At last a battery of the Guard was sent over to support us, but instead of the light artillery, it belonged to the Foot Artillery of the Reserve of 12-pounders. It had the utmost difficulty in moving forward through the mud and only took up position behind us after endless delay. Its first shots were so badly aimed that they blew away a complete troop of our own regiment.

A movement to the rear was ordered. We carried this out at ordinary pace and formed up again behind the battery. The Chasseurs-à-Cheval, Dragoons and Grenadiers-à-Cheval extended their movement further and took position in echelon a short distance behind and to the left of us.

The English cavalry advanced on and off to follow us, but as soon as they came up with our line they stopped, respecting our Lancers above all – the long lances intimidated them. They were limited to firing their pistols at us before retiring behind their infantry line, which made no move. Then a voluntary truce was reached between the combatants due to the complete exhaustion of the troops. Half our squadrons dismounted within musket range.

This suspension of arms lasted about three-quarters of an hour, during which we were hoping that the Emperor's genius would change the face of the battle, forming a general, supported and decisive attack – but nothing! Absolutely nothing!

It was then that we changed from participants to spectators of an incomprehensible drama of which the terrible absurdity was soon recognised, and roundly condemned, by even the least of our simple troopers.

The small plain which we bordered on one side was, as it were, a great circus whose boxes were occupied by the English. Into this bloody arena descended, one after another, poor men destined for death, whose sacrifice was all the easier and quicker since the English, without danger to themselves, were waiting for them at point-blank range.

At first, a few battalions came past our left and presented themselves in column to the English right and its fearless Scots under cover in a wood. When these had been stretched on the ground it was the turn of our Carabinier brigade. This emerged on our right at a gentle trot, crossed the arena alone in column of troops, and rode along all the enemy batteries to attack the English right.

Then Wellington's musketry and batteries awoke together, aiming at the same point, and despite the thunder of their fire, we heard three butcher's cheers. Within a few seconds the Carabiniers had vanished, in death or flight.

To make our grief complete, rumours were running through our ranks: our right was routed; Grouchy had sold himself to the enemy; Bourmont, Clouet, du Barail and many other officers had deserted; a senior officer of the column that had just attacked the Scots had fallen, hit by a case shot, and 200 white cockades [the royalist symbol] had spilled from his shako.

When the English retired from their first position they had left proclamations on the field signed by Louis XVIII which promised pardon, amnesty, retention of rank and post.

When there were no more victims to offer for the great sacrifice, when the circus games had come to an end, our

The Battle of Waterloo

attention was drawn to a new spectacle which worthily crowned that day. On the plateau to our right appeared black lines; they came forward, preceded by their guns. They were the Prussians, who had escaped Grouchy!

Ah! How can I describe to you the consternation among the Guard cavalry? They cried out for the Emperor, whom they had not seen since their commitment to battle – and they would not see him yet!

The order to retreat was given. How ominous was that retreat – a funeral procession.

Our light cavalry brigade, reduced to two and a half squadrons and commanded by Generals Lefebvre-Desnouëttes, Lallemand and Colbert (wounded), retreated slowly and extended its line in order to form a curtain which somewhat concealed our routed army from English observation.

So we marched until we met again on our left what remained of the Old Guard infantry which, with us, formed the extreme rearguard. They were facing to the rear; we halted level with them and turned about too. We numbered at that time 100 to 150 officers and troopers of the Lancers and Chasseurs-à-Cheval, exhausted and wretched.

The sun had almost disappeared, and it was nearly dark.

Our three generals came together in front of our line and some officers joined them – I was a member of this group. A powerful assault column of the enemy was marching on the road and was heading for the Foot Guards' square.

Its head had barely appeared on the crest behind which the Foot Guards were standing, when the latter opened fire. This shooting was well enough co-ordinated, but was perhaps premature; and it seemed to me that it would have produced more effect if it had been delivered from closer to the crest and thus as plunging fire. This fire was answered by a rather poor salvo from the enemy, this was followed by a mêlée which was concealed from me by the night and the distance.

General Lefebvre-Desnouëttes cried out in the greatest excitement that, "It is here that we must all die." He said that no Frenchman could outlive such a horrible day, that we must look for death among this mass of English facing us.

We tried to calm him down. A discussion began in which we all took part and, strangely enough, the man who maintained his sangfroid – who still thought that we had a tomorrow to look forward to, who talked about making a useful retreat all the way to Paris – was the one man who would instantly lose his life if the Bourbons laid hands on him: General Lallemand.[1] He ignored his own interests in order to consider the general situation, discussing matters coolly and with authority.

After several hours we began to make out a muffled noise to our left. This grew louder, and soon we broke out onto the road at Quatre-Bras. Here we came upon the most crowded, breathless and disordered retreat that I ever saw.

We lined up in battle formation, facing to the rear, our right being close to the Charleroi road. This movement was barely completed, when one of our officers said: "There is the Emperor!" At once all eyes turned in the direction of the road and there, among a mass of infantry, vehicles, cavalry and wounded, we saw the Emperor riding, accompanied by two officers wearing greatcoats just like him and followed by four or five Gendarmes d'élite. This was, I believe, at 1 o'clock in the morning.

Recognising troops still under discipline, the Emperor came towards us. Never has such a bright moon lit a more horrible night. The moonlight fell full on the face of the Emperor as he stood in front of our ranks.

Never, even during the retreat from Moscow, had I seen a more confused and unhappy expression on that majestic face.

"Who are you?" asked his majesty. "The Lancers of the Guard." "Ah, yes! The Lancers of the Guard! And where is Piré?"

"Sire, we know nothing of him."

"What, and the 6th Lancers?"

"Sire, we do not know, he was not with us."

"That's right... but Piré?"

"We have no idea," replied General Colbert.

"But who are you?"

"Sire, I am Colbert, and here are the Lancers of your Guard."

---

[1]. Lallemand was condemned to death in the 'White Terror', but reprieved, released and went to the United States where he set up a colony of French refugees in Louisiana. He later returned to Paris, where he died on 4 March 1859.

"Ah, yes... and the 6th Lancers? ... and Piré? Piré?"

One of the generals with him dragged him away, and he disappeared into the night. Our grief knew no bounds.'

Epilogue

In popular mythology, peace descended across Europe on 19 June 1815. In fact, fighting went on until 24 September of that year, when the fortresses of Charlemont and Givet finally surrendered to the Prussians.

Sources

Andolenko, General C.R., *Aigles de Napoleon contre les Drapeaux du Tsar*, Paris, 1969

Bodart, Gaston, *Militär-historisches Kriegs-lexikon*, Vienna and Leipzig, 1908

Chandler, David, *The Campaigns of Napoleon*, London, 1966

Esposito, V.J., and Elting, J.R., *An Atlas of the Napoleonic Wars*, London, 1999

Martinien, A., *Tableaux pars Corps et pars Batailles des Officiers Tués et Blessés Pendant les Guerres de l'Empire 1805–1815*, Paris, 1890

Pawley, Roland, *The Red Lancers*, 1998

Siborne, Captain W., *History of the Waterloo Campaign*, London, 1990

Six, Georges, *Dictionnaire Biographique des Generaux & Amiraux Français de la Revolution et de l'Empire (1792–1814)*, Paris, 1934

Appendices
Orders of Battle

Abbreviations
The following abbreviations may be found throughout the orders of battle

ChL = Chevaux-Légers, light horse (often lancers)
DBdeLi = Demi-Brigade d'Infanterie de Ligne; equivalent of a Line Infantry Regt.
DBdeLe = Demi-Brigade d'Infanterie Légère; equivalent of a Light Infantry Regt.
Drag = Dragoon
EH = Erzherzog (Grand Duke)
FML = Feldmarschall-Leutnant; an Austrian rank
FZM = Feldzugmeister; an Austrian rank
GdB = Général de brigade
GdD = Général de division
GdK = General der Kavallerie
GLt = General-Leutnant
GM = General-Major
GrR = Grenadier Regiment
HA = Horse Artillery
HAB = Horse Artillery battery
IR = Infantry Regiment
KGL = King's German Legion
KüR = Cuirassier regiment
MG = Major-General
MR = Musketeer Regiment
NCR = National Cavalry Regiment
RFA = Regiment of Field Artillery
RHA = Regiment of Horse Artillery
UR = Ulan (Lancer) Regiment

Author's Note
The many sources consulted in the preparation of these orders of battle are frequently contradictory in many aspects, particularly with regard to numbers of troops. Some originals contain obvious errors (for example where sub-totals do not add up to the grand totals given) which can now no longer be resolved without a lifetime of careful research on each point of detail.

The Battle of Marengo

The French Army   **Napoleon Bonaparte, First Consul**

Consular Guard
- Grenadiers-à-Pied   2 battalions
- Grenadiers-à-Cheval   2 squadrons
- Chasseurs-à-Cheval   2 squadrons
- Artillery   1 HAB   6 guns

Armée de la Reserve   *General Alexandre Berthier*
Chief of Staff   *Lt-Gen Pierre Dupont de Letang*

Victor's Corps   *Lt-Gen Claude Victor*

Gardanne's Division   *GdD Gaspard Gardanne*
- 44th DBdeLi   3 battalions
- 101st DBdeLi   3 battalions
- 102nd DBdeLi   1 company

Chambarlhac's Division   *GdB Jacques Chambarlhac de Laubespin*
GdB Herbin-Dassaux, GdB Olivier Rivaud
- 24th DBdeLe   3 battalions
- 43rd DBdeLi   3 battalions
- 96th DBdeLi   3 battalions

Artillery
- 4th HAB, 5th Regt HA   4 guns
- 10th HAB, 6th Regt HA   3 guns
- 1st Battalion, Train

Lannes's Corps   *GdD Jean Lannes*
GdB Joseph Mainoni
- 5th Dragoons   2 squadrons
- 28th DBdeLi   3 battalions

Watrin's Division   *GdD François Watrin*
GdB Claude Gency
- 6th DbdeLe   3 battalions

GdB Jean Malher
- 22nd DBdeLi   3 battalions
- 40th DBdeLi   3 battalions

Artillery
- 2nd HAB, 2nd Regt HA   3 guns, 1 how

Desaix's Corps   *Lt-Gen Louis Desaix*

Monnier's Division   *GdD Jean Monnier*
GdB Jean Schilt
- 19th DBdeLe   2 battalions

GdB Claude Carra St. Cyr
- 70th DBdeLi   3 battalions
- 72nd DBdeLi   3 battalions

Artillery   2 guns

Boudet's Division   *GdD Jean Boudet*
GdB Louis Musnier
- 9th DbdeLe   3 battalions

GdB Louis Guenand
- 30th DBdeLi   2 battalions
- 59th DBdeLi   3 battalions

Orders of Battle

Cavalry
    3rd Cavalry   4 squadrons
    4th Hussars   4 squadrons
Artillery
    4th HAB, 2nd Regt HA   6 guns

Cavalry   **Lt-Gen Joachim Murat**
GdB François-Étienne Kellermann
    2nd Cavalry Regt   4 squadrons
    20th Cavalry Regt   4 squadrons
    21st Cavalry Regt   4 squadrons

GdB Pierre Champeaux
    1st Dragoons   2 squadrons
    8th Dragoons   4 squadrons
    9th Dragoons   2 squadrons

GdB Jean Rivaud
    11th Hussars   4 squadrons
    12th Hussars   4 squadrons
    21st Chasseurs-à-Cheval   4 squadrons

GdB Bernard Duvigneau*
    6th Dragoons   4 squadrons
    12th Chasseurs-à-Cheval   4 squadrons

Moncey's Division   **Lt-Gen Bon-Adrien Moncey**
GdD Jean Lapoype
    1st DbdeLe   2 battalions
    29th DBdeLi   3 battalions
    91st DBdeLi   2 battalions

* Duvigneau did not command at Marengo as he had been injured in a fall from his horse the previous day

The Austrian Army   GdK Michael Melas

Chief of Staff   **GM Anton Zach**

Right Column   **FML Andreas Graf von O'Reilly**
GM Rousseau

5th Hussar Regiment	2 squadrons	230 men
8th Hussar Regiment Nauendorff	2 squadrons	300 men
Grenz-IR Deutsch-Banat Nr 12	4th Battalion	533 men
Grenz-IR Warasdiner-Kreuzer Nr 5	1st Battalion	755 men
Grenz-IR Otocaner Nr 2	3rd Battalion	298 men
Grenz-IR Oguliner Nr 3	1st Battalion	602 men
8th Light Drag Regt Württemberg	1 squadron	113 men
Artillery	1 HAB	6 guns

Main Column   **FML Karl, Graf Haddick von Futak**
Advanced Guard   **Oberst Johann Frimont von Palota**

3rd Lt Inf Battalion Bach		277 men
4th Lt Inf Battalion Am Ende		291 men
1st Light Drag Regt Kaiser	2 squadrons	312 men
Jäger-zu-Pferde Regt Bussy	2 squadrons	186 men
Pioneers	1 company	115 men
Artillery	1 HAB	6 guns

The Battle of Marengo

Division of   **FML Karl, Graf Haddick von Futak**
GM Giovanni Pilatti

1st Light Drag Regt Kaiser	3 squadrons	460 men
4th Light Drag Regt Karaczay	6 squadrons	1,053 men

GM F Bellegarde

IR EH Anton Nr 52	2 battalions	855 men
IR Jellacic Nr 53	2 battalions	956 men
Artillery		8 x 3-pdr guns

Division of   **FML Konrad Freiherr Kaim**
GM de Briey

IR Franz Kinsky Nr 47	2⅓ battalions	1,640 men
Artillery		4 x 3-pdr guns

GM Knesevich

IR EH Ferdinand Nr 23	3 battalions	2,188 men
Artillery		4 x 3-pdr, 1 x 6-pdr guns

GM La Marseille

IR EH Joseph Nr 63	3 battalions	1,111 men
Artillery		4 x 3-pdr, 1 x 6-pdr guns

Reserve   **GM Franz Graf St. Julien**

IR Michael Wallis Nr 11	3 battalions	2,209 men
Artillery		6 x 3-pdr guns

Grenadier Battalions   **FML Morzin**
GM Christof Freiherr von Lattermann

Kleinmayer Battalion	378 men
Paar Battalion	357 men
St. Julien Battalion	580 men
Schiaffinati Battalion	408 men
Weber Battalion	393 men
Artillery	10 x 3-pdr guns

GM Weidenfeld

Gorschen Battalion		291 men
Khevenhüller Battalion		384 men
Pers Battalion		293 men
Pertusi Battalion		555 men
Piret Battalion		226 men
Weissenwolf Battalion		491 men
Pioneers	4 companies	575 men
Artillery		12 x 3-pdr guns

Reserve Cavalry   **FML Anton Freiherr von Elsnitz**
GM Nobili

3rd Light Drag Regt EH Johann	6 squadrons	859 men
9th Light Drag Regt Liechtenstein	6 squadrons	1,059 men

GM Nimbsch

7th Hussar Regiment	8 squadrons	1,353 men
9th Hussar Regiment Erdoedy	6 squadrons	988 men

Reserve Artillery   5 x HABs   30 guns

Orders of Battle

Left Column — *FML Karl Freiherr von Ott*
Advanced Guard — *GM Gottesheim*

Mariassy Jägers	4 companies	245 men
IR Fröhlich Nr 28	1 battalion	523 men
10th Light Drag Regt Lobkowitz	2 squadrons	245 men
Artillery	1 x HAB	6 guns

Division of — *FML Schellenberg*
GM Retz

IR Fröhlich Nr 28	2 battalions	1,046 men
IR Mittrowsky Nr 40	3 battalions	853 men
Pioneers	1 company	115 men
Artillery	6 x 3-pdr, 2 x 6-pdr, 2 x 12-pdr guns	

GM Sticker

IR Splenyi Nr 51	2 battalions	737 men
IR J Colloredo Nr 57	3 battalions	1,369 men
10th Light Drag Regt Lobkowitz	4 squadrons	495 men
Artillery	8 x 3-pdr, 1 x 6-pdr guns	

Division of — *FML Ludwig Freiherr von Vogelsang*
GM Ulm

IR Hohenlohe Nr 17	2 battalions	912 men
IR Stuart Nr 18	3 battalions	1,282 men
Artillery	6 x 3-pdr, 1 x 6-pdr gun	

Technical Troops — *Maj Graf Hardegg*

Pontoniers	1 company	60 men
Sappers	1 company	91 men
Miners	2 companies	96 men

On Outpost Duties

At Acqui

1st Light Drag Regt Kaiser	1 squadron	45 men

At Asti

Grenz-IR Deutsch-Banater Nr 12	1 battalion	482 men
Light Infantry Battalion Rohan	1 battalion	578 men

At Casale

2nd Hussar Regiment EH Joseph	4 squadrons	1,097 men

At Novi

8th Hussar Regiment Nauendorff	1 squadron	126 men

The Battle of Austerlitz

French Order of Battle for Austerlitz
The following order of battle is a combination of that given by Scott Bowden in *Napoleon and Austerlitz* (which is shown in normal type) and that given by Adrian Pascal in his *Histoire de l'Armée* (which is shown in this condensed/*condensed italic* font). The totals shown as being present for each regiment vary between the sources in many cases. Pascal also gives the names of most of the regimental commanders and the total of casualties suffered by each regiment. Where a colonel's name is marked with an asterisk, this indicates that no matching name has been found in Six. Where total strengths for corps or similar formations are given as a number of men this includes artillery train, engineers and similar personnel who are not normally listed separately for reasons of space.

Le Grande Armée	Emperor Napoleon I		
Chief of the General Staff	***Marshal Berthier***		
Grand Marshal of the Palace	*GdD Duroc*		
Military Administration	*Intendant General Daru*		
Commander of the Artillery	*GdD Songis*		
Commander of the Engineers	*GdD Marescot*		
The Imperial Guard	***Marshal Bessières***		
Chief of Staff	*GdB Roussel*		
Commander of the Artillery	*Col Couin*		
1st Infantry Brigade	*GdB Hulin*		
Grenadiers-à-pied *Dorsenne*		2 battalions	1,519 men
2nd Brigade	*GdB Soules*		
Chasseurs-à-pied *Groux*		2 battalions	1,613 men
3rd Brigade (The Italian Guard)	*Col Lecchi*		
Foot Grenadiers		1 battalion	
Foot Chasseurs		1 battalion	
		Total Italian Guard	753 men
1st Cavalry Brigade	*GdB Ordener*		
Grenadiers-à-Cheval		4 squadrons	706 men
		600 men, 2 killed, 22 wounded	
2nd Cavalry Brigade	*Col Morland*		
Chasseurs-à-Cheval		4 squadrons	375 men
		276 men, 19 killed, 65 wounded	
Mamelukes		½ squadron	48 men
3rd Brigade	*GdD Savary*		
Detached to Brünn			
Gendarmerie d'Élite		2 squadrons,	? men
Artillery		24 guns	283 men
1st HA Company, 2nd HA Company, HA Company of the Italian Guard			
	Total	Guard	7,367 men
			3,320 horses
			24 guns

The Reserve (1st Div, V Corps) *GdD Duroc* (commanding in place of Oudinot, wounded)
This division was formed of combined grenadier regiments, drawn from the elite companies of various regiments as shown below

1st Brigade	*GdB Laplanche-Morthières*	
1st Grenadier Regiment	2 battalions	
1st Battalion (coys from 13th Line IR)		340 men
2nd Battalion (coys from 58th Line IR)		422 men

Orders of Battle

2nd Grenadier Regiment	2 battalions	
1st Battalion (coys from 9th Line IR)		499 men
2nd Battalion (coys from 81st Line IR)		526 men
2nd Brigade	***GdB Dupas***	
3rd Grenadier Regiment	2 battalions	
1st Battalion (coys from 2nd Light IR)		399 men
2nd Battalion (coys from 3rd Light IR)		542 men
4th Grenadier Regiment	2 battalions	
1st Battalion (coys from 28th Light IR)		425 men
2nd Battalion (coys from 31st Light IR)		432 men
3rd Brigade	***GdB Ruffin***	
5th Grenadier Regiment	2 battalions	
1st Battalion (coys from 12th Light IR)		538 men
2nd Battalion (coys from 15th Light IR)		532 men
Artillery		
1st Company, 1st RFA	4 x 8-pdr, 2 x 4-pdr	
Part of 4th Company, 5th RHA	2 x 8-pdr	
Total	1st Division	5,016 men
		8 guns
		328 horses

I Corps — *Marshal Bernadotte*
Chief of Staff — ***GdD Victor-Léopold Berthier***
Commander of the Artillery — ***GdD Jean-Baptiste Eblé***
Commander of the Engineers — ***Col Morio***

Advanced Guard
Detached from Drouet's 2nd Division

1st Brigade	***GdB Frere***	
27th Light IR *Col Chanelot*	3 battalions	2,069 men
		2,069 men, 4 killed, 57 wounded

1st Division — ***GdD Rivaud (Rivaux) de la Raffinière***

1st Brigade	***GdB Dumoulin***	
8th Line IR *Col Autie*	3 battalions	1,858 men
		1,858 men, 2 wounded
2nd Brigade	***GdB Pacthod***	
45th Line IR *Col Barrie*	3 battalions	1,603 men
		1,603 men, 3 killed, 8 wounded
54th Line IR *Col Philippon*	3 battalions	1,614 men
		1,614 men, 1 wounded
Artillery		
1st Company, 8th RFA	4 x 3-pdr, 1 how	
2nd Company, 3rd RHA	4 x 6-pdr, 1 how	
		111 men

2nd Division — ***GdD Drouet d'Erlon***
1st Brigade
Detached as Corps Advanced Guard

2nd Brigade	***GdB Werle***	
94th Line IR *Col Razout*	3 battalions	1,814 men
		1,814 men, 11 killed, 101 wounded
95th Line IR *Col Pescheux*	3 battalions	1,903 men
		1,903 men, 2 killed, 27 wounded
Artillery		
2nd Company, 8th RFA	5 x 3-pdr, 1 x how	
3rd Company, 3rd RHA	5 x 6-pdr, 1 x how	
		118 men

The Battle of Austerlitz

Light Cavalry Division   **GdD Kellermann**
Detached to the Cavalry Reserve

I Corps' Artillery Reserve
Detached to serve with General Wrede's Bavarians

	Total	I Corps	11,346 men
			22 guns

III Corps   **Marshal Davout**
Chief of Staff   *GdB Fournier de Loysonville Daultane*
Commander of the Artillery   *GdD Sorbier*
Commander of the Engineers   *GdB Andreossy*

1st Infantry Division   ***GdD Caffarelli du Falga***
Detached to V Corps

2nd Infantry Division   ***GdD Friant***
1st Brigade   *GdB Kister*
    15th Light IR *Col Desailly*   2 battalions   300 men
(less 6 companies detached to 2nd Batt, 5th Combined Gren Regt in V Corps)
                                            754 men, 31 killed, 146 wounded

    33rd Line IR *Col Saint-Remond*   2 battalions   500 men
                                            1,214 men, 17 killed, 300 wounded

2nd Brigade   *GdB Lochet*
    48th Line IR *Col Barbanegre*   2 battalions   800 men
                                            1,365 men, 17 killed, 125 wounded

    111th Line IR *Col Gay*   2 battalions   700 men
                                            1,440 men, 10 killed, 121 wounded

3rd Brigade   *GdB Heudelet*
    108th Line IR *Col Higgonet*   2 battalions   800 men
    15th Light IR   Voltigeur company   70 men
                                            1,637 men, 148 killed, 316 wounded

Artillery
    2nd Company, 7th RFA   4 x 8-pdr, 2 x how
    1st Company, 5th RHA   2 x 8-pdr, 1 how
                                                            161 men, 8 wounded

3rd Division   ***GdD Gudin***
Not present for the battle

Light Cavalry Division   ***GdB Viallanes***
Not present for the battle

1st Dragoon Division   ***GdD Klein***
Attached from the Reserve Cavalry

1st Brigade   *GdB Senerols*   total 574 men
    1st Dragoons   3 squadrons
    2nd Dragoons   3 squadrons
2nd Brigade   *GdB Lasalle*   total 602 men
    4th Dragoons   3 squadrons
    20th Dragoons   3 squadrons
3rd Brigade   *GdB Milet*   total 617 men
    14th Dragoons   3 squadrons
    26th Dragoons   3 squadrons

Orders of Battle

4th Dragoon Division — *GdD Bourcier*
Attached from the Reserve Cavalry
  1st Dragoons — 3 squadrons — 120 men
1st Brigade — *GdB Sahuc*
  15th Dragoons *Col Barthélemi* — 3 squadrons — 90 men
      385 men
  17th Dragoons *Col Saint-Dizier** — 3 squadrons — 120 men
      474 men
2nd Brigade — *GdB Laplanche*
  18th Dragoons *Col Lefebvre* — 3 squadrons — 150 men
      553 men
  19th Dragoons *Col Caulaincourt* — 3 squadrons — 160 men
      408 men
3rd Brigade — *GdB Verdière*
  25th Dragoons *Col Rigeaud** — 3 squadrons — 100 men
      396 men
  27th Dragoons *Col Fereye** — 3 squadrons — 90 men
Artillery
  3rd Company, 2nd RHA — 2 x 8-pdr guns, 1 how
    Artillery and Train — 81 men

    Total   III Corps   3,107 infantry
                               2,623 cavalry
                               12 guns

IV Corps — *Marshal Soult*
Chief of Staff — *GdD Saligny*
Commander of the Artillery — *GdB Lariboissière*
Commander of the Engineers — *Col Poitevin*

1st Division — *GdD Saint-Hilaire*
1st Brigade — *GdB Morand*
  10th Light IR *Col Pouzet* — 2 battalions — 1,488 men
    1,488 men, 50 killed, 279 wounded
2nd Brigade — *GdB Thiebault*
  14th Line IR *Col Mazas* — 2 battalions — 1,551 men
    1,551 men, 22 killed, 219 wounded
  36th Line IR *Col Houdart-Lamotte* — 2 battalions — 1,643 men
    1,486 men, 30 killed, 373 wounded
3rd Brigade — *GdB Varé*
  43rd Line IR *Col Vivies* — 2 battalions — 1,598 men
    1,598 men, 42 killed, 421 wounded
  55th Line IR *Col Ledru* — 2 battalions — 1,658 men
    1,709 men, 44 killed, 306 wounded
Artillery
  12th Company, 5th RFA — 2 x 8-pdr, — 98 men
                         2 x 4-pdr, 2 x how
  16th Company, 5th RFA — 2 x 8-pdr — 22 men
    120 Artillery men in all, 1 killed, 11 wounded

2nd Division — *GdD Vandamme*
1st Brigade — *GdB Schiner*
  24th Light IR *Col Pourailly* — 2 battalions — 1,291 men
    1,291 men, 126 killed, 364 wounded
2nd Brigade — *GdB Fery* Ferey, which is correct
  4th Line IR *Col Bigarre* — 2 battalions — 1,658 men
    1,658 men, 18 killed, 193 wounded
  28th Line IR *Col Edighoffen* — 2 battalions — 1,599 men
    1,599 men, 9 killed, 75 wounded

The Battle of Austerlitz

3rd Brigade	*GdB Candras*		
46th Line IR *Col Latrille*		2 battalions	1,350 men
		1,350 men, 20 killed, 288 wounded	
57th Line IR *Col Rey*		2 battalions	1,734 men
		1,743 men, 6 killed, 75 wounded	
Artillery			
13th Company, 5th RFA		2 x 8-pdr,	95 men
		2 x 4-pdr, 2 x how	
16th Company, 5th RFA		2 x 8-pdr	22 men
		117 artillerymen in all, 1 killed, 11 wounded	
3rd Division	***GdD Legrand***		
1st Brigade	*GdB Merle*		
26th Light IR *Col Pouget*		2 battalions	1,564 men
		1,564 men, 54 killed, 57 wounded	
Tirailleurs du Po *Col Hulot*		1 battalion	340 men
		308 men, 29 killed, 154 wounded	
Tirailleurs Corses		1 battalion	519 men
Not shown by Pascal			
2nd Brigade	*GdB Fery*		
3rd Line IR *Col Schobert*		3 battalions	1,644 men
		1,644 men, 56 killed, 376 wounded	
3rd Brigade	*GdB Lavasseur*		
18th Line IR *Col Ravier*		2 battalions	1,402 men
		1,402 men, 30 killed, 102 wounded,	
75th Line IR *Col Lhuillier de Hoff*		2 battalions	1,688 men
		1,688 men, 15 killed, 82 wounded	
Artillery			
14th Company, 5th RFA		2 x 8-pdr,	94 men
		2 x 4-pdr, 2 x how	
16th Company, 5th RFA		2 x 8-pdr	22 men
	Artillery and Train	220 men, 1 killed, 11 wounded	
1st Light Cavalry Brigade	*GdB Margaron*		
8th Hussars *Col Franceschi*		3 squadrons	359 men
		341 men, 2 killed, 19 wounded	
11th Chasseurs-à-Cheval *Col Bessières*		4 squadrons	343 men
		308 men, 10 killed, 49 wounded,	
26th Chasseurs-à-Cheval *Col Digeon*		3 squadrons	316 men
		256 men, 1 killed, 10 wounded	
Artillery			
4th Company, 5th RHA		5 x 8-pdr	91 men
3rd Dragoon Division	***GdD Beaumont***		
Attached from the Reserve Cavalry			
1st Brigade	***GdB Boye*** *Boyer (Boye is correct according to Six)*		
5th Dragoons *Col Lacour**		3 squadrons	234 men
		264 men, 5 killed, 29 wounded	
8th Dragoons *Col Becker**		3 squadrons	289 men
		295 men, 11 killed, 24 wounded	
2nd Brigade	***GdB Scalfort***		
9th Dragoons		3 squadrons	297 men
		292 men, 5 killed, 8 wounded	
12th Dragoons *Col Pages*		3 squadrons	291 men
		312 men, 13 killed, 18 wounded	

Orders of Battle

3rd Brigade	*Adjt Cmdt Devaux*		
16th Dragoons *Col Clement*		3 squadrons	242 men
		258 men, 21 killed, 30 wounded	
21st Dragoons		3 squadrons	285 men
Not shown in this brigade by Pascal			

Artillery
3rd Company, 2nd RHA		2 x 8-pdr, 1 how	42 men
	Artillery and Train	88 men, 1 killed, 2 wounded	

IV Corps Artillery Reserve
17th and 18th Companies, 5th RFA		6 x 12-pdr	154 men
Total	IV Corps		22,727 infantry
			2,656 cavalry
			38 guns

V Corps — *Marshal Lannes*
Chief of staff — *GdB Compans*
Commander of the Artillery — *GdB Foucher de Careil*
Commander of the Engineers — *Col Kirgener*

1st Division — *GdD Duroc*
Detached to the Army Reserve

2nd Division — *GdD Gazan*
In Vienna with Marshal Mortier

3rd Division — *GdD Suchet*
1st Brigade — *GdB Claparède*

17th Light IR *Col Vedel*		2 battalions	1,373 men
		1,412 men, 9 killed, 30 wounded	

2nd Brigade — *GdB Beker*

34th Line IR *Col Dumoustier*		3 battalions	1,615 men
		1,812 men, 19 killed, 178 wounded	
40th Line IR *Col Legendre de Harvesse*		2 battalions	1,149 men
		1,428 men, 24 killed, 255 wounded	

3rd Brigade — *GdB Valhubert* Killed in action

64th Line IR *Col Nerin**		2 battalions	1,052 men
		1,140 men, 15 killed 73 wounded	
88th Line IR *Col Curial*		2 battalions	1,428 men
		1,596 men, 46 killed, 122 wounded	

1st Infantry Division — *GdD Caffarelli du Falga*
Attached from III Corps
1st Brigade — *GdB Demont* Wounded in action

17th Line IR *Col Conroux*		2 battalions	1,561 men
		1,661 men, 25 killed, 75 wounded	
30th Line IR *Col Walther**		2 battalions	1,011 men
		1,372 men, 23 killed, 338 wounded	

2nd Brigade — *GdB De Billy* Adjt-Cmdt Coehord

51st Line IR *Col Bonnet*		2 battalions	1,214 men
		1,278 men, 15 killed, 49 wounded	
61st Line IR *Col Nicolas*		2 battalions	1,175 men
		1,206 men, 9 killed, 22 wounded	

3rd Brigade — *GdB Eppler*

13th Light IR *Col Castex**		2 battalions	1,240 men
		1,407 men, 25 killed, 141 wounded	

Artillery
1st Company, 7th RFA		4 x 8-pdr, 2 x how	77 men
		89 men, 1 killed, 11 wounded	

The Battle of Austerlitz

Light Cavalry Division   **GdB Fauconnet**
Detached to the Reserve Cavalry

2nd Dragoon Division   **GdD Walther**
Attached from the Reserve Cavalry

1st Brigade   GdB Sebastiani de la Porta
Adjt-Cmdt Lacroix (Sebastiani is correct)

3rd Dragoons *Col Fiteau*	3 squadrons		177 men
		193 men, 5 killed, 11 wounded	
6th Dragoons *Col Lebaron**	3 squadrons		150 men
		164 men, 2 killed, 12 wounded	

2nd Brigade   GdB Roget de Belloguet

10th Dragoons *Col Cavaignac*	3 squadrons		207 men
		246 men, 1 killed, 16 wounded	
11th Dragoons *Col Bourdon**	3 squadrons		196 men
		214 men, 18 wounded	

3rd Brigade   GdB Boussart *Boussard (Boussart is correct)*

13th Dragoons *Col Debroc*	3 squadrons		269 men
		271 men, 2 killed)	
22nd Dragoons *Col Carrie*	3 squadrons		134 men
		168 men, 7 killed, 26 wounded	

Artillery
2nd Company, 2nd RHA   2 x 8-pdr, 1 how   41 men
Artillery and Train   96 men, 1 killed, 6 wounded

Corps Artillery

15th Company, 5th RFA	2 x 12-pdr, 4 x 8-pdr, 2 x 4-pdr	61 men
16th Company, 5th RFA	2 x 8-pdr	22 men
5th Company, 1st RFA	2 x 8-pdr, 2 x how	28 men

121 artillerymen, 3 killed, 7 wounded

Total   V Corps   12,818 infantry
                             1,133 cavalry
                             23 guns

Reserve Cavalry   **Marshal Murat**
Chief of Staff   **GdD Belliard**
Commander of the Artillery   **GdB Hanicque**
Commander of the Engineers   Hayelle

1st Heavy Cavalry Division   **GdD Nansouty**
1st Brigade   GdB Piston

1st Carabiniers *Col Cochois*	3 squadrons		205 men
		205 men, 2 killed, 24 wounded	
2nd Carabiniers *Col Morin*	3 squadrons		181 men
		181 men, 17 wounded	

2nd Brigade   GdB La Houssaye

2nd Cuirassiers *Col Yvendorff*	3 squadrons		304 men
		1 killed, 16 wounded	
9th Cuirassiers *Col Doumerc*	3 squadrons		280 men
		280 men, 9 wounded	

3rd Brigade   GdB Saint-Germaine

3rd Cuirassiers *Col Preval*	3 squadrons		333 men
		333 men, 44 killed, 27 wounded	
12th Cuirassiers *Col Belfort*	3 squadrons		277 men
		277 men, 3 killed, 1 wounded	

Artillery
4th Company, 2nd RHA   92 men

Orders of Battle

2nd Heavy Cavalry Division — **GdD d'Hautpoul**
1st Brigade — **Col Noirot** Adjt-Cmdt Fontaine
    1st Cuirassiers Col Guiton   3 squadrons   388 men
                                                      388 men, 11 killed, 27 wounded
    5th Cuirassiers Col Noireau   3 squadrons   375 men
                                                      375 men, 21 killed, 14 wounded
2nd Brigade — **GdB Saint-Sulpice**
    10th Cuirassiers Col Lataye   3 squadrons   254 men
                                                    254 men, 3 killed, 14 wounded
    11th Cuirassiers Col Fouler   3 squadrons   327 men
                                                    327 men, 14 killed, 37 wounded
Artillery
    4th Company, 2nd RHA   2 x 8-pdr, 1 how   42 men
                                                                                  42 men

1st Dragoon Division — **GdD Klein**
Detached to III Corps

2nd Dragoon Division — **GdD Walther**
Detached to V Corps

3rd Dragoon Division — **GdD Beaumont**
Detached to IV Corps

4th Dragoon Division — **GdD Bourcier**
Detached to III Corps

Light Cavalry Brigade — **GdB Milhaud**
    16th Chasseurs-à-Cheval Col Durosnel   3 squadrons   338 men
                                                        320 men, 55 wounded
    22nd Chasseurs-à-Cheval Col Latour-Maubourg   3 squadrons   272 men
                                                        234 men, 4 wounded

1st Light Cavalry Division — **GdD Kellermann**
Attached from I Corps
1st Brigade — **GdB van Marizy** Marizi (Marizy is correct)
    2nd Hussars Col Barbier   3 squadrons   328 men
                                                      328 men, 8 killed, 32 wounded
    5th Hussars   3 squadrons   342 men
    Also shown by Pascal as part of this brigade
        5th Chasseurs-à-Cheval Col Corbineau   317 men, 31 killed, 58 wounded
        5th Hussars Col Schwartz   342 men, 8 killed, 15 wounded
2nd Brigade — **GdB Picard**
Pascal shows the 2nd and 4th Hussars brigaded together under Picard
    4th Hussars Col Burthe   3 squadrons   280 men
                                                     280 men, 16 killed, 47 wounded
    5th Chasseurs-à-Cheval Col Corbineau   3 squadrons   317 men
Artillery
    1st Company, 3rd RHA   2 x 6-pdr   79 men
                                         2 x 3-pdr
                                         2 x how

Light Cavalry Division — **GdB Fauconnet**
Attached from V Corps
1st Brigade — **GdB Trellard**
    9th Hussars Col Guyot   3 squadrons   233 men
                                                     227 men, 22 wounded
    10th Hussars Col Beaumont   3 squadrons   261 men
                                                     179 men, 1 killed, 8 wounded

The Battle of Austerlitz

2nd Brigade   *GdB Fauconnet*
  13th Chasseurs-à-Cheval   3 squadrons   259 men
  21st Chasseurs-à-Cheval   3 squadrons   256 men
         Total   Reserve Cavalry   5,810 cavalry
                                                                                                                                                                                12 guns

Grand Artillery Park
  16th, 17th and 18th Companies, 7th RFA   230 men
  18 Austrian 3-pdr

Grand Totals
  Scott Bowden   58,155 infantry, 11,558 cavalry, 157 guns
  Pascal   68,867 men: 54,171 infantry, 12,200 cavalry, 2,406 artillery and train
             Losses: 1,288 killed, 6,903 wounded

The Allied Order of Battle for the Battle of Austerlitz

All Austrian regiments and commanders are marked with an asterisk *.
As well as the officers listed Czar Alexander and Kaiser Franz were also present.

Commander-in-Chief   **General Count Kutuzov**
Chief of staff   *GM Weyrother*
ADCs to the Czar   *Gen Adjt Dolgoruki*
                         Gen Adjt Prince Volkonski
                         Gen Adjt Winzingerode

Austrian Commander   *FML Prince Johann von Liechtenstein* *
GHQ Escort
  KürR Kaiser Nr 1   2 squadrons
  (from 3rd Division)

The Allied army was in effect divided into the following corps: Buxhowden (Advanced Guard and 1st, 2nd and 3rd Columns); Kollowrat (4th Column); Liechtenstein (Cavalry); Bagration (Right Wing); Constantine (Imperial Guard).

Corps of   *Gen of Infantry Count Buxhowden*
Advanced Guard   *FML Baron Kienmayer*
Brigade of   *GM Carneville* *
  Wiener Jäger*   2 companies
  Broder Grenz-IR Nr 7*   1 battalion
  1st Szeckler Grenz-IR Nr 14*   2 battalions
  2nd Szeckler Grenz-IR Nr 15*   2 battalions
Brigade of   *GM Stutterheim* *
  ChL Regt O'Reilly Nr 3*   8 squadrons
  Ulan Regt Merveldt Nr 1*   ¼ squadron
  1st Austrian HAB*   6 guns
Brigade of   *GM Graf Nostitz* *
  Ulan Regt Schwarzenberg Nr 2*   ½ squadron
  Hussar Regt Hesse-Homburg Nr 4*   8 squadrons

Orders of Battle

Brigade of	*GM Prince Moriz Liechtenstein**	
Hussar Regt Szeckler Nr 11*	6 squadrons	
Sisozev III Cossacks	5 squadrons	
Melentiev III Cossacks	5 squadrons	
2nd Austrian HAB*		6 guns
Total	Advanced Guard	6,340 men
		2,940 horses
		12 guns

1st Column	*Lt-Gen Dmitri Docturov*	
Denisov Cossacks	5 squadrons	
Vasilisissoiev Cossacks	5 squadrons	
Brigade of	*MG Lewis*	
7th Jägers	1st Battalion	
New Ingermanland MR	3 battalions	
Yaroslav MR	3 battalions	
Brigade of	*MG Alexander Urusov*	
Briansk MR	3 battalions	
Vladimir MR	3 battalions	
3rd Brigade		
Kiev GrR	3 battalions	
Moscow MR	3 battalions	
Viatka MR	3 battalions	
Artillery		
	2 x HABs	12 guns each
Total	1st Column	13,730 infantry
		2,000 cavalry
		24 heavy guns
		52 regimental guns

2nd Column	*Lt-Gen Count Louis-Alexandre Langeron*	
Brigade of	*MG Zachari Olsuviev I*	
4th Jägers	2 battalions	
Perm MR	3 battalions	
Vyborg MR	3 battalions	
Brigade of	*MG Count Sergei Kaminski I*	
Fanagoria GrR	3 battalions	
Kursk MR	3 battalions	
Riask MR	3 battalions	
Total	2nd Column	10,840 men
		30 regimental guns

3rd Column	*Lt-Gen Ishami Przbychevski*	
Brigade of	*MG Müller III*	
7th Jägers	2nd and 3rd Battalions	
Galizin MR	3 battalions	
Brigade of	*MG Strik*	
Butyrsk MR	3 battalions	
Narva MR	3 battalions	
Brigade of	*MG Loshakov*	
8th Jägers	1st Battalion (remnants)	
Azov MR	3 battalions	
Podolsk MR	3 battalions	
Demisod Cossacks	5 squadrons	
Total	3rd Column	5,448 men
		500 horses
		30 regimental guns

The Battle of Austerlitz

4th Column   *FZM Vincent Count Kollowrat**
Division of   *Lt-Gen Mikhail Miloradovich*
Brigade of   *MG Vodnianski*
  Drag Regt EH Johann Nr 1*   2 squadrons
Brigade of   *MG Grigori Berg I*
  Little Russia GrR   3 battalions
  Novgorod MR   3 battalions
Brigade of   *MG Sergei Repninski II*
  Apcheron MR   3 battalions
  Smolensk MR   3 battalions
Artillery Reserve

           Total   6,965 men
                  125 cavalry
                  24 heavy guns
                  52 light guns

Division of   *FZM Vincent Count Kollowrat**
Brigade of   *GM Rottermund**
  IR Auersperg Nr 24*   6th Battalion
  IR Kaunitz Nr 20*   6th Battalion
  IR Salzburg Nr 23*   6 battalions
Brigade of   *GM Jurscheck**
  IR Beaulieu Nr 58*   3rd Battalion
  IR Czartorisky Nr 9*   6th Battalion
  IR Kaiser Nr 1*   6th Battalion
  IR Kerpen Nr 49*   6th Battalion
  IR Lindenau Nr 29*   4th Battalion
  IR Württemberg Nr 38*   3rd Battalion
Artillery   1 x Russian FAB   24 guns
         2 x Austrian 12-pdr FAB

          Total   12,099 men
                    24 guns

Cavalry Corps (5th Column) *FML Prince Johann von Liechtenstein**
Division of   *FM Prince Hohenlohe**
Brigade of   *GM Caramely**
  KürR Lothringen Nr 7*   8 squadrons
  KürR Nassau Nr 5*   8 squadrons
Brigade of   *GM Weber**
  KürR Kaiser Nr 1*   6 squadrons
         Total   1,100 men

Artillery
  1 x Austrian HAB   4 x 6-pdr guns, 2 x how

Division of   *GL Essen II*
Brigade of   *MG Dmitri Czepelev*
MG Baron Mellor-Sackomelski II
  Guard Cossacks   5 squadrons
  Lancer Regt Grand Duke Constantine   10 squadrons
  St. Petersburg Drag Regt   5 squadrons
  Lifeguard Cuirassier Regt   5 squadrons

         Total   2,500 men

Orders of Battle

Division of — *Gen-Adjt Fedor Uvarov*
Brigade of — *MG Penitzki*
  Kharkov Drag Regt — 5 squadrons
  Czernigov Drag Regt — 5 squadrons
  Elizabethgrad Hussar Regt — 10 squadrons
Artillery
  1 x Austrian HAB
  1 x Russian FAB
  ½ Russian HAB
Cossacks
  Sotnia of Denisov — 5 squadrons
  Sotnia of Gordeev — 5 squadrons
  Sotnia of Issazev — 5 squadrons

    Total — 2,000 men
            24 guns

Corps of the Right Wing — *Lt-Gen Prince Bagration*
Brigade of — *MG Dolguruki*
  5th Jägers — 3 battalions
  6th Jägers — remnants of 3 battalions
Brigade of — *MG Kaminski II*
  Archangel MR — 3 battalions
Brigade of — *MG Engelhardt*
  Old Ingermanland MR — 3 battalions
  Pskov MR — 3 battalions
Brigade of — *MG Wittgenstein*
  Mariupol Hussar Regt — 10 squadrons
  Pavlograd Hussar Regt — 10 squadrons
Brigade of — *MG Voropaitzki*
  Cuirassier Regt 'His Majesty' — 5 squadrons
  Tver Drag Regt — 5 squadrons
  St. Petersburg Drag Regt — 5 squadrons
Cossacks — *MG Chaplitz*
  Kissilev Cossacks — 5 squadrons
  Malachov Cossacks — 5 squadrons
  Chaznenkov Cossacks — 5 squadrons
Artillery
    18 battalion guns
  1 x Russian HAB◊
  2 x Austrian HABs◊
  ◊*Arrived from Olmütz in the afternoon of the battle*

    Total — 8,500 infantry
           3,500 cavalry
           30 guns initially
           42 later

Russian Imperial Guard — *Gen of Cavalry Grand Duke Constantine*
Division of — *Lt-Gen Andrei Kollogribov*
MG Jankovich
  Life Guard Cossacks — 2 squadrons
  Life Guard Hussars — 5 squadrons
MG Depreradovich II
  Horse Guards — 5 squadrons
  Chevalier Guard — 5 squadrons

The Battle of Austerlitz

Division of   *Lt-Gen Miliutin*

MG Depreradovich I
- Life Guard Jägers — 1 battalion
- Semenovski Footguards — 2 battalions
- Preobrazhenski Footguards — 2 battalions
- Ismailovski Footguards — 2 battalions

MG Lubanov
- Life Grenadiers — 3 battalions

Artillery
- 20 battalion guns (6-pdrs and 10-pdr unicorns)
- one Guard HAB — 10 guns
- one Guard FAB — 10 guns

    Total   Imperial Guard   6,730 infantry
                                                  3,430 cavalry
                                                  40 guns

Total   ***Allied armies***   50,850 infantry and pioneers
                                                  11,799 regular cavalry
                                                  2,340 Cossacks
                                                  318 guns

The Battle of Preussisch-Eylau

La Grande Armée   **Emperor Napoleon I**

The Imperial Guard Infantry   *Marshal Lefebvre*
1st Grenadiers-à-Pied *GdB Hulin*	2 battalions
1st Chasseurs-à-Pied *GdB Soules*	2 battalions
Foot Artillery	1 battery

The Imperial Guard Cavalry   *Marshal Bessières*
Grenadiers-à-Cheval *GdD Walther*	4 squadrons
Chasseurs-à-Cheval *Col-Maj Dahlmann*	4 squadrons
Mamelukes	1 squadron
Horse Artillery	1 battery

III Corps   *Marshal Davout*
This corps arrived on the field at midday on 8 February

1st Division   *GdD Morand*
GdB De Billy, GdB Brouard
13th Light IR	2 battalions	
17th Line IR	2 battalions	
30th Line IR	2 battalions	
Artillery	2 x FAB	13 guns

2nd Division   *GdD Friant*
GdB Kister, GdB Lochet, GdB Grandeau
33rd Line IR	2 battalions	
48th Line IR	2 battalions	
108th Line IR	2 battalions	
111th Line IR	2 battalions	
Artillery	1x FAB	8 guns

3rd Division   *GdD Gudin*
GdB Petit, GdB Gautier, GdB ?
12th Line IR	2 battalions	
25th Line IR	2 battalions	
85th Line IR	2 battalions	
21st Line IR	3 battalions	
Artillery	one FAB	8 guns

Light Cavalry Brigade   *GdB Viallanes*
1st Chasseurs	3 squadrons
12th Chasseurs	3 squadrons

Artillery Reserve	3 x FAB, 1 x HAB	17 guns

IV Corps   *Marshal Soult*

1st Division   *GdD Saint-Hilaire*
GdB Candras, GdB Varé
10th Light IR	2 battalions	
35th Line IR	2 battalions	
43rd Line IR	2 battalions	
55th Line IR	2 battalions	
Artillery	2 x FAB	12 guns

The Battle of Preussisch-Eylau

2nd Division   *GdD Leval*
GdB Ferey, GdB Schiner, GdB Vivies
- 24th Light IR — 2 battalions
- 4th Line IR — 2 battalions
- 28th Line IR — 2 battalions
- 46th Line IR — 2 battalions
- 57th Line IR — 2 battalions
- Artillery — 2 x FAB — 12 guns

3rd Division   *GdD Legrand*
GdB Des Essarts, GdB Lavasseur
- 26th Light IR — 2 battalions
- 18th Line IR — 2 battalions
- 75th Line IR — 2 battalions
- Tirailleurs Corses — 1 battalion
- Tirailleurs du Po — 1 battalion
- Artillery — 2 x FAB — 12 guns

Artillery Reserve — 1 x FAB — 6 guns

VI Corps   *Marshal Ney*
This corps did not arrive on the field until 6:00 p.m.

1st Division   *GdD Marchand*
GdB Roguet, GdB Villatte
- 6th Light IR — 2 battalions
- 39th Line IR — 2 battalions
- 69th Line IR — 2 battalions
- 76th Line IR — 2 battalions
- Artillery — 1 x FAB

2nd Division   *GdD Gardanne*
GdB ?, GdB ?, GdB ?
- 25th Light IR — 2 battalions
- 27th Line IR — 2 battalions
- 50th Line IR — 2 battalions
- 59th Line IR — 2 battalions
- Artillery — 1 x FAB

Light Cavalry Brigade   *GdB Colbert*
- 10th Chasseurs — 3 squadrons
- 9th Hussars — 3 squadrons
- Artillery — 1 x HAB

Artillery Reserve — 1 x FAB, 1 x HAB

VII Corps   *Marshal Augereau*

1st Division   *GdD Desjardin*
GdB Lapisse, GdB Lefranc
- 16th Light IR — 4 battalions
- 44th Line IR — 4 battalions
- 105th Line IR — 4 battalions
- 14th Line IR — 2nd Battalion
- Artillery — 1 x FAB, 1 x HAB — 8 guns

2nd Division   *GdD Heudelet de Bierre*
GdB Amey, GdB ?, GdB Sarrut
- 7th Light IR — 3 battalions
- 24th Line IR — 3 battalions
- 63rd Line IR — 2 battalions
- Artillery — 1 x FAB, 1 x HAB — 8 guns

Light Cavalry Brigade	*GdB Durosnel*		
20th Chasseurs		3 squadrons	
Artillery		1 x HAB	4 guns
Artillery Reserve		2 x FAB	16 guns

Reserve Cavalry — *Marshal Murat*

1st Cuirassier Division — *GdD Nansouty*
9th Cuirassiers	4 squadrons	
11th Cuirassiers	4 squadrons	
Artillery	½ HAB	3 guns

2nd Cuirassier Division — *GdD d'Hautpoul*
GdB Saint-Sulpice, GdB Verdière
1st Cuirassiers	4 squadrons	
5th Cuirassiers	4 squadrons	
10th Cuirassiers	4 squadrons	
Artillery	½ HAB	3 guns

1st Dragoon Division — *GdD Klein*
GdB Ferenolz, GdB Lamotte, GdB Picard
1st Dragoons	3 squadrons	
2nd Dragoons	3 squadrons	
4th Dragoons	4 squadrons	
14th Dragoons	4 squadrons	
22nd Dragoons	3 squadrons	
23rd Dragoons	3 squadrons	
Artillery	½ HAB	3 guns

2nd Dragoon Division — *GdD Grouchy*
GdB Boussart, GdB Milet, GdB Roget
3rd Dragoons	3 squadrons	
4th Dragoons	3 squadrons	
6th Dragoons	3 squadrons	
10th Dragoons	3 squadrons	
11th Dragoons	3 squadrons	
Artillery	½ HAB	3 guns

3rd Dragoon Division — *GdD Milhaud*
GdB Boye, GdB Latour-Maubourg, GdB Marizy
5th Dragoons	4 squadrons	
8th Dragoons	4 squadrons	
12th Dragoons	4 squadrons	
19th Dragoons	4 squadrons	
21st Dragoons	4 squadrons	
Artillery	½ HAB	3 guns

4th Dragoon Division — *GdD Sahuc*
GdB ?, GdB Laplanche, GdB Margaron
17th Dragoons	3 squadrons	
18th Dragoons	3 squadrons	
19th Dragoons	3 squadrons	
27th Dragoons	3 squadrons	
Artillery	½ HAB	3 guns

The Battle of Preussisch-Eylau

Light Cavalry Division   *GdD Lasalle*
GdB Milhaud
  1st Hussars   3 squadrons
  3rd Hussars   3 squadrons
GdB Bruyères
  5th Hussars   3 squadrons
  13th Chasseurs   3 squadrons
GdB Watier
  11th Chasseurs   4 squadrons
  1st Bavarian ChL Regt Kronprinz   4 squadrons
Artillery   1 x HAB   8 guns

      Total   **French Army**   90 battalions
                                            131 squadrons
                                            150 guns
                                            ca 75,000 men

The Allied Army   **General of Cavalry Baron Bennigsen**
2nd Division   *Lt-Gen Count Ostermann-Tolstoi*
Brigade of   *Maj-Gen Koschin*
  Lifeguard Cuirassier Regiment   5 squadrons
  Kargopol Dragoons   5 squadrons
  Isum Hussars   10 squadrons
  Cossacks Pulks of Illovaiski IX, Zhefremov III
Brigade of   *Maj-Gen Mazofskoi*
  Pavlov Grenadiers   3 battalions
  Rostov MR   3 battalions
Brigade of   *Maj-Gen Sukin II*
  Petersburg Grenadiers   3 battalions
  Jeletz MR   3 battalions
Brigade of   *Maj-Gen Count Lieven*
  1st Jägers   3 battalions
  20th Jägers   3 battalions

Artillery
  2 x 12-pdr, 2 x 6-pdr and 1 x HAB   5 batteries

3rd Division   *Lt-Gen Baron Sacken*
Brigade of   *Maj-Gen Count Piotr Pahlen*
  Little Russia Cuirassiers   5 squadrons
  Kurland Dragoons   5 squadrons
  Soum Hussars   10 squadrons
  Cossack Pulks of Illovaiski X, Papuzin
Brigade of   *Maj-Gen Uschakov*
  Tauride Grenadiers   3 battalions
  Lithuanian MR   3 battalions
Brigade of   *Maj-Gen Titov II*
  Kaporski MR   3 battalions
  Muromsk MR   3 battalions
Brigade of   *Maj-Gen Netting*
  Chernigov MR   3 battalions
  Dnieproff MR   3 battalions

Artillery
  2 x 12-pdr, 3 x 6pdr and 1 x HAB   6 batteries

4th Division — **Lt-Gen Prince Galitzin**
(Lt Gen Somov commanded in the battle)

Brigade of — Maj-Gen Baron Korff
  St. George Cuirassiers — 5 squadrons
  Pskov Dragoons — 5 squadrons
  Polish Cavalry Regiment — 10 squadrons
  Cossack Pulks of Grekov IX and Grekov XVIII

Brigade of — Maj-Gen Somov
  Tenguinsk MR — 3 battalions
  Tula MR — 3 battalions

Brigade of — Maj-Gen Arseniev
  Tobolsk MR — 3 battalions

Brigade of — Maj-Gen Barclay de Tolly
  Kostroma MR — 3 battalions
  Polotsk MR — 3 battalions
  3rd Jägers — 3 battalions

Artillery
  2 x 12-pdr, 3 x 6-pdr and 1 x HAB — 6 batteries

5th Division — **Lt-Gen Tuchkov I**

Brigade of — Maj-Gen ?
  Sievsk MR — 3 battalions
  Kaluga MR — 3 battalions

Brigade of — Maj-Gen ?
  Mohilev MR — 3 battalions
  Perm MR — 3 battalions

Brigade of — Maj-Gen Sacken
  Riga Dragoons — 5 squadrons
  Kazan Dragoons — 5 squadrons
  Elizabethgrad Hussars — 10 squadrons
  Lithuanian Cavalry Regiment — 5 squadrons
  Cossack Pulk of Gordezov

Artillery
  2 x 12-pdr, 2 x 6-pdr, 1 x HAB — 5 batteries

7th Division — **Lt-Gen Docturov**

Brigade of — Maj-Gen Markov
  Ekaterinoslav Grenadiers — 3 battalions
  Moscow MR — 3 battalions

Brigade of — Maj-Gen ?
  Vladimir MR — 3 battalions
  Voronezh MR — 3 battalions

Brigade of — Maj-Gen ?
  Azov MR — 3 battalions
  Pskov MR — 3 battalions
  5th Jägers — 3 battalions

Brigade of — Maj-Gen Chaplitz
  Moscow Dragoons — 5 squadrons
  Ingermanland Dragoons — 5 squadrons
  Pavlograd Hussars — 10 squadrons
  Cossack Pulks of Malachov and Andronov

Artillery
  2 x 12-pdr, 3 x 6-pdr and 1 x HAB — 6 batteries

The Battle of Preussisch-Eylau

8th Division   *Lt-Gen Essen III*
MG Zapolski, Col Count Pahlen
- Moscow Grenadiers   3 battalions
- Schlusselburg MR   3 battalions
- Old Ingermanland MR   3 battalions
- Podolsk MR   3 battalions
- Archangelgorod MR   3 battalions
- 7th Jägers   3 battalions
- St. Petersburg Dragoons   5 squadrons
- Livland Dragoons   5 squadrons
- Olviopol Hussars   10 squadrons
- Cossack Pulks of Kieselev and Sysoyev

Artillery
- 4 x 12-pdr and 1 x HAB   5 batteries

14th Division   *Maj-Gen Kamenskoi*
Lt-Gen Anrepp had been killed at Möhrungen on 25 January
The brigade structure of this division is not known
- Bielosersk MR   3 battalions
- Ryasansk MR   3 battalions
- Sophia MR   3 battalions
- Ouglitz MR   3 battalions
- 23rd Jägers   3 battalions
- 26th Jägers   3 battalions
- Finland Dragoons
- Mittau Dragoons
- Grodno Hussars   20 squadrons in all

     1 x 12-pdr battery
     2 x 6-pdr batteries

     Total   **Russian Army**   123 battalions
     145 squadrons
     11 Cossack Pulks
     ca 300 guns
     ca 58,000 men

Prussian Corps   *Gen-Lt Anton Wilhelm von l'Estocq*

Avantgarde
- Auer Dragoons Nr 6   10 squadrons
- Towarczys   ½ squadron
     1 x HAB

1st Division   *GM von Diericke*
- Fabecky Grenadiers   1 battalion
- IR von Rüchel Nr 2   2 battalions
- Baczko Dragoons Nr 7   5 squadrons
- Kurassiers von Wagenfeld Nr 4   4 squadrons
     ½ x HAB

2nd Division   *GM von Rembow*
- Vac von Schlieffen Grenadiers   1 battalion
- IR von Schöning Nr 11   2 battalions

3rd Division
  Towarczys
  Fusiliers
  Wyborg MR (Russian)
  21st Jägers (Russian)

General von Auer
9½ squadrons
2 companies
3 battalions
3 battalions
½ x HAB

Total	*Prussian Corps*	12½ battalions
		29 squadrons
		9 guns
		ca 5,584 men

The Battle of Albuera

Allied Forces — **Marshal William Carr Beresford**

O = officers, M = men

	Present O	Present M	Killed O	Killed M	Wounded O	Wounded M	Missing O	Missing M	Total O+M
2nd Division — *General Sir William Stewart*									
Colborne's Brigade									
1st Battalion 3rd Foot	27	728	4	212	14	234	2	177	643
2nd Battalion 31st Foot	20	398	–	29	7	119	–	–	155
2nd Battalion 48th Foot	29	423	4	44	10	86	9	190	343
2nd Battalion 66th Foot	24	417	3	52	12	104	–	101	272
Hoghton's Brigade									
29th Foot	31	476	5	75	12	233	–	11	336
1st Battalion 48th Foot	33	464	3	64	13	194	–	6	280
1st Battalion 57th Foot	31	616	2	87	21	318	–	–	428
Abercrombie's Brigade									
2nd Battalion 28th Foot	28	491	–	27	6	131	–	–	164
2nd Battalion 34th Foot	28	568	3	30	4	91	–	–	128
2nd Battalion 39th Foot	33	449	1	14	4	77	–	2	98
Light Troops									
5th Battalion 60th Foot	4	142	–	2	1	18	–	–	21
2nd Division totals	288	5,172	25	636	104	1,605	11	487	2,868
4th Division — *General Sir Galbraith Cole*									
Myers's Brigade									
1st Batt. 7th Fusiliers	27	687	–	65	15	277	–	–	357
2nd Batt. 7th Fusiliers	28	540	2	47	13	287	–	–	349
1st Batt. 23rd Fusiliers	41	692	2	74	11	246	–	6	339
Kemmis's Brigade *(detachment)*									
One company each of 2/27th, 1/40th, 97th Foot	8	157	1	5	–	14	–	–	20
4th Division totals	104	2,076	5	191	39	824	–	6	1,065
Independent Brigade — *General Sir Karl Alten*									
1st Light Battalion KGL	23	565	–	4	4	59	–	2	69
2nd Light Battalion KGL	19	491	1	3	1	31	–	1	37
KGL Brigade totals	42	1,056	1	7	5	90	–	3	106
Cavalry — *General Sir William Lumley*									
De Grey's Brigade									
3rd Dragoon Guards	23	351	1	9	–	9	–	1	20
4th Dragoons	30	357	–	3	2	18	2	2	27
13th Light Dragoons	23	380	–	–	–	1	–	–	1
Cavalry totals	76	1,088	1	12	2	28	2	3	48
Artillery									
British (2 batteries)	9	246	–	3	1	10	–	1	15
KGL (2 batteries)	10	282	–	–	1	17	1	30	49
Artillery totals	19	528	–	3	2	27	1	31	64
Staff officers			1		7				8
British totals	529	9,920	33	849	159	2,574	14	530	4,159

Portuguese Troops
Totals are shown as all ranks together

	Present	Killed	Wounded	Missing	Total
Harvey's Brigade *(4th Division)*					
11th Line IR	1,154	2	6	5	13
23rd Line IR	1,201	4	15	–	19
1st Batt Loyal Lusit. Legion	572	66	95	10	171
Hamilton's Division					
2nd Line IR	1,225	3	5		8
14th Line IR	1,204	–	2		2
4th Line IR	1,271	9	51		60
10th Line IR	1,119	–	11	–	11
Collin's Brigade					
5th Line IR	985	10	40	10	60
5th Caçadores	400	5	25	1	31
Otway's Cavalry					
1st Regiment	327	–	–	–	–
7th Regiment	314	–	2	–	2
5th Regiment	104	–	–	–	–
8th Regiment	104	–	–	–	–
Artillery (2 batteries)	221	2	8	–	10
Staff officers		1	1		2
Portuguese totals	10,201	102	261	26	389

Spanish Army of the Centre *General Joachim Blake*
All infantry regiments are of one battalion except where noted.

	Present		Killed		Wounded		Total
	O	M	O	M	O	M	
Vanguard Division	*General Manuel Lardizabal*						
Campo Mayor, Canarias, 2nd León, Murcia (2 batts)	107	2,291	4	59	13	215	291
3rd Division	*General Fransisco Ballasteros*						
Barbastro, Cangas de Tineo Castropol, 1st Catalonia, Infiesto, Lena, Pavia	154	3,371	3	64	15	193	275
4th Division	*General José Zayas*						
Ciudad Real, Irlanda, Legion Estranjera Patria, Spanish Guards 2nd and 4th batts Toledo, Walloon Gds 4th Batt	197	4,685	–	106	26	549	681
Cavalry	*General Loy*						
Escuadrón de Instrucción, Granaderos, Husares de Castilla, Santiago	93	1,072	–	7	2	31	40
Artillery (1 battery)	7	96	–	2	–	7	9
Staff officers				2		9	11
Totals for Blake's force	558	11,515	9	238	65	995	1,307

The Battle of Albuera

Det. from Army of Galicia — General Carlos de España

	Present		Killed		Wounded		Total
	O	M	O	M	O	M	
Infantry							
Voluntarios de Navarra, Rey, Zamora, 1 company of Sappers	57	1,721	–	–	4	29	33
Cavalry	*Maj-Gen Penne Villemur*						
7 squadrons, various regts	87	634	–	11	3	14	28
Artillery (1 battery)	4	58	–	–	–	–	–
Totals for de España's force	148	2,413	–	11	7	43	61

Grand totals Allied Army

	Present	Losses
British	10,449	4,159
Portuguese	10,201	389
Spaniards	14,634	1,368
Total	35,284	5,916

The French Army — Marshal Soult, Duke of Dalmatia

	Present		Killed		Wounded		Missing		Total
	O	M	O	M	O	M	O	M	O+M
V Corps									
1st Division — *General Girard*									
34th Line IR, 2nd and 3rd Battalions	23	930	4	104	13	298	–	–	19
40th Line IR, 1st and 2nd Battalions	35	778	4	35	9	226	1	73	348
64th Line IR, 1st, 2nd, 3rd Battalions	50	1,539	5	99	18	361	1	168	651
88th Light IR, 2nd and 3rd Battalions	21	878	–	–	5	253	6	141	405
2nd Division — *General Count Gazan*									
21st Line IR, 2nd and 3rd Battalions	43	745	3	61	11	154	2	24	255
100th Line IR, 1st and 2nd Battalions	33	705	4	50	8	152	2	51	267
28th Line IR, 1st, 2nd, 3rd Battalions	62	1,305	7	53	10	313	1	112	496
103rd Line IR, 1st, 2nd, 3rd Battalions	38	1,252	4	48	10	148	3	74	287
V Corps totals	305	8,132	31	450	84	1,905	15	643	3,128
Brigade of — *General Werle*									
12th Line IR, 1st, 2nd, 3rd Battalions	62	2,102	3	108	14	511	1	132	769
55th Line IR, 1st, 2nd, 3rd Battalions	58	1,757	4	68	6	235	–	38	351
58th Line IR, 1st, 2nd, 3rd Battalions	55	1,587	6	23	15	258	2	24	328

Orders of Battle

	Present		Killed		Wounded		Missing		Total
	O	M	O	M	O	M	O	M	O+M
Brigade of	*General Godinot*								
16th Light IR									
1st, 2nd, 3rd Battalions	49	1,624	2	39	7	321		12	381
51st Line IR									
1st, 2nd, 3rd Battalions	65	2,186		2	–	1		–	3
Combined grenadiers (11 coys) of 45th, 63rd, 95th Line IR of I Corps and 4th Polish IR of IV Corps	33	1,000	?	?	?	?	?	?	372
Total infantry	627	18,388	46	690	126	3,230	18	849	5,332
Cavalry	*General Viscount Latour–Maubourg*								
Brigade of	*General Baron Briche*								
2nd Hussars	23	282	1	4	3	57	–	8	73
10th Hussars	24	238	1	3	4	21	–	3	32
21st Chasseurs	21	235	–	3	3	19	–	–	25
Brigade of	*General Baron Bron de Bailly*								
4th Dragoons	21	385	3	27	1	38	–	1	70
20th Dragoons	22	244	1	6	3	10	1	4	25
26th Dragoons	27	394	1	5	2	12	–	1	21
Brigade of	*General Baron Bouvier des Eclaz*								
14th Dragoons	17	299	–	6	1	17	–	–	24
17th Dragoons	17	297	–	12	3	29	–	1	45
27th Dragoons	14	235	–	2	3	11	–	3	19
Unattached cavalry									
1st Vistula Leg. Lancers	28	563	1	41	9	78	1	–	130
27th Chasseurs	22	409	–	7	2	11	1	5	26
4th Spanish Chasseurs	14	181	–	2	–	4	–	–	6
Total cavalry	250	3,762	8	118	34	307	3	26	496
Artillery	18	590	1	19	3	72	–	–	95
Other units	25	600	?	?	?	?	?	?	?
Staff officers			5		8				13
French Army totals	920	23,340	60	827	171	3,610	21	875	5,936

The French casualty figures are incomplete due to the missing returns. Soult, for example, reported only 262 officer casualties, but Martinien's lists show at least 362.

The Clash at García Hernandez

French Forces *General Foy*
1st Division of the Army of Portugal
  6th Light IR   2 battalions
  39th Line IR   2 battalions
  69th Line IR   2 battalions
  76th Line IR   2 battalions
  1 x battery of foot artillery

Light Cavalry *General Curto*
  13th Chasseurs-à-Cheval
  15th Chasseurs-à-Cheval
  22nd Chasseurs-à-Cheval
  26th Chasseurs-à-Cheval
  28th Chasseurs-à-Cheval
  3rd Hussars.
                  Total   4,000 men

All had suffered in the battle the previous day and their morale was shaky at best.

    Blackwood's Magazine of March 1913 states that there were elements of 10 French cavalry units present. They are listed as 6th, 11th and 25th Dragoons, 13th, 14th, 15th, and 31st Chasseurs, 1st Hussars, Lancers of Berg and the 1st Legion de Gendarmerie. This conflicts with the order of battle of the French Army given by Oman. He lists the 15th Dragoons not the 15th Chasseurs, and does not show the Lancers of Berg nor the Gendarmerie. He does, however, show an Escadron de Marche, which might have contained men of several different regiments.

Allied Forces

Brigade of   *Maj-Gen George Anson*
  11th Light dragoons   390 all ranks
  12th Light Dragoons*   330 all ranks
  16th Light dragoons   270 all ranks
Brigade of   *Maj-Gen Eberhardt von Bock*
  1st Dragoons KGL   360 all ranks
  2nd Dragoons KGL   400 all ranks

* *possibly not present at this action*

Borodino – The Northern Flank

Russian Forces

I Cavalry Corps   *Lt-Gen F.P. Uvarov*
Maj-Gen Chalikov
  Life Guard Hussars
  Life Guard Ulans
  Life Guard Dragoons
Maj-Gen Chernich
  Niezine Dragoons
  Elizabethgrad Hussars
  Horse Artillery Battery Nr 5.
                                      Total   20 squadrons, 12 guns
                                                            about 2,500 men

Cossacks   *Gen of Cavalry Platov*
Cossack pulks of Ilovaiski V, Grekov XVIII, Chariton VII, Denisov VII, Zhirov, Vlasov III, the Simferopol Tartar Regiment and part of the Ataman's pulk
                                        Total   about 5,500 men

Bogdanovich gives a slightly different Russian order of battle for this raid: 'Uvarov's command was: Life Guard Hussars, Life Guard Ulans, Life Guard Dragoons, Life Guard Cossacks, Rjaschin [*sic*] Dragoons, Elizabethgrad Hussars, and Horse Artillery Battery Nr 2.' However, his version of the composition of Platov's Cossacks is the same.

French and Allied Forces

IV Corps   *Prince Eugène, Viceroy of Italy*
The Italian Royal Guard   *General Lecchi*
  Honour Guard Regt *Col Bataglia*   5 companies
  Velites *Col Moroni*   2 battalions
  Infantry Regiment *Maj Crovi*   2 battalions
  Conscript Regiment *Col Peraldi*   2 battalions
  Marines   1 company

Cavalry   *General Triaire*
  Guard Dragoons *Col Jaquet*   4 squadrons
  Queen's Dragoons *Col Narboni*   4 squadrons

Artillery of the Italian Guard
  1 x FAB, 1 x HAB

13th Division   *General Delzons*
Brigade of   *Gen Huard*
  8th Light IR   2 battalions
  84th Line IR   4 battalions
Brigade of   *Gen Roussel*
  92nd Line IR   4 battalions
  1st Croatian IR *Col Slivarich*   2 battalions
Brigade of   *Gen Guyon*
  106th Line IR   4 battalions
Artillery
  1 x FAB, 1 x HAB

Borodino – The Northern Flank

14th Division **General Broussier**
Brigade of *Gen de Sivray*
  18th Light IR 2 battalions
  9th Line IR 4 battalions

Brigade of *Gen Alméras*
  35th Line IR 4 battalions
  Spanish IR Joseph Napoleon *Maj Doreille* 2 battalions

Brigade of *Gen Pastol*
  53rd Line IR 4 battalions

Artillery
  1 x FAB, 1 x HAB

15th (Italian) Division **General Pinot**
Reached the battlefield only at 9 p.m.
  1st Light IR 1 battalion
  3rd Light IR 4 battalions
  2nd Line IR 4 battalions
  3rd Line IR 4 battalions
  Dalmatian IR 3 battalions

Artillery
  1 x FAB, 1 x HAB

Attached Light Cavalry **General Ornano**
12th Light Cavalry Brigade *Gen Ferrière*
  9th Chasseurs-à-Cheval 4 squadrons
  19th Chasseurs-à-Cheval 4 squadrons

13th Light Cavalry Brigade *Gen Villata*
  2nd (Italian) Chasseurs-à-Cheval 4 squadrons
  3rd (Italian) Chasseurs-à-Cheval 4 squadrons

21st (Bavarian) Light Cav Brig *Gen Count von Seydewitz*
  3rd Bavarian Chevauxlégers 4 squadrons
  6th Bavarian Chevauxlégers 4 squadrons

22nd (Bavarian) Light Cav Brig *Gen Preysing*
  4th Bavarian Chevauxlégers 4 squadrons
  5th Bavarian Chevauxlégers 4 squadrons

Corps Reserve Artillery
  4 x FAB, 4 x HAB

Borodino – The Grand Battery

French and Allied Forces

II Reserve Cavalry Corps   *Gen Montbrun*

2nd Light Cavalry Division   *Gen Sebastiani*
7th Light Brigade   *Gen St. Genies*
  11th Chasseurs   4 squadrons
  12th Chasseurs   4 squadrons

8th Brigade   *Gen Baurth*
  5th Hussars   4 squadrons
  9th Hussars   4 squadrons

16th Brigade   *Gen Subervie*
  1st Prussian Ulans   4 squadrons
  3rd Württemberg Jäger-zu-Pferde   4 squadrons
  10th Polish Hussars   4 squadrons

2nd Cuirassier Division   *Gen Watier*
1st and 2nd Brigades   *Gen Caulaincourt*
  5th Cuirassiers   4 squadrons
  8th Cuirassiers   4 squadrons

3rd Brigade   *Gen Richter*
  10th Cuirassiers   4 squadrons
  2nd Chasseurs   4 squadrons
Artillery
  2 HA batteries

4th Cuirassier Division   *Gen Defrance*
1st Brigade   *Gen Berckheim*
  1st Carabiniers   4 squadrons

2nd Brigade   *Gen L'Eritage*
  2nd Carabiniers   4squadrons

3rd Brigade   *Gen ?*
  1st Cuirassiers   4 squadrons
  4th Chevaux-légers   4 squadrons
Artillery
  2 HA batteries

III Reserve Cavalry Corps   *Gen Grouchy*

3rd Light Cavalry Division   *Gen Chastel*
10th Brigade   *Gen Gauthrin*
  6th Chasseurs   4 squadrons
  8th Chasseurs   4 squadrons

11th Brigade   *Gen Gerard*
  6th Hussars   4 squadrons
  25th Chasseurs   4 squadrons

17th Brigade   *Gen Dommanget*
  1st Bavarian Chevauxlégers   4 squadrons
  2nd Bavarian Chevauxlégers   4 squadrons
  Saxon Chevauxlégers 'Prinz Albrecht'   4 squadrons

Borodino – The Grand Battery

6th Heavy Cavalry Division   *Gen La Houssaye*
1st Brigade   *Gen Thiry*
  7th Dragoons   4 squadrons
  23rd Dragoons   4 squadrons
2nd Brigade   *Gen Seron*
  28th Dragoons   4 squadrons
  30th Dragoons   4 squadrons
Artillery   2 HA batteries

IV Reserve Cavalry Corps   *Gen Latour-Maubourg*
4th (Polish) Light Cav Division   *Gen Rozniecki*
29th Brigade   *Gen Turno*
  3rd Lancers   4 squadrons
  11th Lancers   4 squadrons
  16th Lancers   4 squadrons
  17th Lancers   4 squadrons
Artillery   2 HA batteries

7th Cuirassier Division   *Gen Lorge*
1st Brigade   *Gen Thielemann*
  Saxon Garde du Corps   4 squadrons
  Saxon Kürassier-Regiment von Zastrow   4 squadrons
  14th Polish Cuirassiers   4 squadrons
2nd Brigade   *Gen Lepel*
  1st Westphalian Kürassiers   4 squadrons
  2nd Westphalian Kürassiers   4 squadrons
Artillery   2 HA batteries

Russian Forces

The Right Wing   *General of Infantry Miloradovich*

IV Infantry Corps   *Lt-Gen Count Ostermann-Tolstoi*
11th Infantry Division   *MG Bachmetiev II*
  Kexholm Infantry Regiment
  Pernov Infantry Regiment
  Polotsk Infantry Regiment
  Jeletz Infantry Regiment
  1st Jägers
  33rd Jägers
  Combined Grenadier Regiment
  Position Battery Nr 11

23rd Infantry Division   *MG Bachmetiev I*
  Rilsk Infantry Regiment
  Katherinenburg Infantry Regiment
  Seleginsk Infantry Regiment
  18th Jägers
  1 Combined Grenadier Battalion
  Light Battery Nr 44

                      Total IV Corps   23 battalions, 2 batteries
                                                          9,500 men, 24 guns

II Cavalry Corps — *General Adjutant Baron Korf*

Pskov Dragoon Regiment	4 squadrons
Moscow Dragoon Regiment	4 squadrons
Kargopol Dragoon Regiment	4 squadrons
Ingermanland Dragoon Regiment	4 squadrons
Polish Ulan Regiment	8 squadrons
Isum Hussars	8 squadrons
Horse Artillery Battery Nr 4	

Total II Cavalry Corps 32 squadrons, 1 battery
3,500 men, 12 guns

The Centre — *General of Infantry Docturov*

VI Infantry Corps — *General of Infantry Docturov*

7th Infantry Division — *Lt-Gen Kapsevich*
Moscow Infantry Regiment
Pskov Infantry Regiment
Sophia Infantry Regiment
Libau Infantry Regiment
11th Jägers
36th Jägers
Combined Grenadier Regiment
Position Battery Nr 7

24th Infantry Division — *MG Likhachev*
Ufa Infantry Regiment
Schirvan Infantry Regiment
Butyrsk Infantry Regiment
Tomsk Infantry Regiment
19th Jägers
40th Jägers
Combined Grenadier Regiment
Position Battery Nr 24

Total VI Infantry Corps: 28 battalions, 2 batteries
9,900 men, 24 guns

III Cavalry Corps — *General Adjutant Baron Korf**
MG Kreuz

Kurland Dragoon Regiment	4 squadrons
Orenburg Dragoon Regiment	4 squadrons
Siberia Dragoon Regiment	4 squadrons
Irkutsk Dragoon Regiment	4 squadrons
Soum Hussars	8 squadrons
Mariupol Hussars	8 squadrons
Horse Artillery Battery Nr 9	

Total III Cavalry Corps 32 squadrons, 1 battery
3,700 men, 12 guns

*(Due to the sickness of Lt-Gen Count Piotr Pahlen)

Borodino – The Grand Battery

The Reserve of the Right Wing and The Centre

V Infantry Corps   *GD Constantine*
Lifeguard Division   *Lt-Gen Lavrov*

Preobrazhenski Guard Regiment	3 battalions
Semenovski Guard Regiment	3 battalions
Izmailovski Guard Regiment	3 battalions
Lithuania Guard Regiment	3 battalions
Finland Guard Regiment	3 battalions
Guard Jäger Regiment	3 battalions
Guards Marine Equipage	1 battalion

Grenadier Division   *MG Kantakuzen*
Combined Grenadier Battalions of the
4th Div, 17th Div, 1st Div and 3rd Div   2 battalions each

        Total V Infantry Corps   27 battalions, 13,000 men

1st Cuirassier Division   *MG Borosdin II*

Chevalier Guards	4 squadrons
Life Horse Guards	4 squadrons
Cuirassier Regiment 'His Majesty'	4 squadrons
Cuirassier Regiment 'Her Majesty'	4 squadrons
Astrakhan Cuirassiers	4 squadrons

        Total 1st Cuirassier Div   20 squadrons, 2,400 men

Reserve Artillery   *MG Count Kutaisov*
        then *MG Loewenstern* and *Colonel Eiler*

Guards Artillery Brigade	6 batteries
Position Artillery	6 batteries
Light Artillery	9 batteries
Horse Artillery	5 batteries

        Total Artillery Reserve   8,400 men
        26 batteries, 300 guns

The 2 horse artillery batteries of the Guard had 8 guns each, the 1st Light Battery of the Guard included 2 guns of the Marine Equipage.

The Left Wing   *General of Infantry Prince Bagration*
        Lt-Gens Prince Gorshakov II, Prince Galitzin

VII Infantry Corps   *Lt-Gen Raevski*
26th Infantry Division   *MG Paskevich*
Ladoga Infantry Regiment
Poltava Infantry Regiment
Nizhgorod Infantry Regiment
Orel Infantry Regiment
5th Jägers
42nd Jägers
Position Battery Nr 26, Light Battery Nr 47

12th Infantry Division   *MG Vasilchikov*
Narva Infantry Regiment
Smolensk Infantry Regiment
Alexopol Infantry Regiment
New Ingermanland Infantry Regiment
6th Jägers
41st Jägers

        Total VII Infantry Corps: 24 battalions, 2 batteries
        10,800 men, 24 guns

VIII Infantry Corps — *Lt-Gen Borosdin II*
2nd Grenadier Division — *MG Prince Karl von Mecklenburg*
Kiev Grenadier Regiment
Astrakhan Grenadier Regiment
Moscow Grenadier Regiment
Fanagoria Grenadier Regiment
Siberia Grenadier Regiment
Little Russia Grenadier Regiment
Position Battery Nr 2, Light Battery Nr 3

27th Infantry Division — *MG Neverovski*
Vilna Infantry Regiment
Simbirsk Infantry Regiment
Odessa Infantry Regiment
Tarnopol Infantry Regiment
49th Jägers
50th Jägers

	Total VIII Infantry Corps	24 battalions, 2 batteries
		11,200 men, 24 guns

IV Cavalry Corps — *MG Count Sievers I*
Kharkov Dragoon Regiment	4 squadrons
Czernigov Dragoon Regiment	4 squadrons
Kiev Dragoon Regiment	4 squadrons
New Russia Dragoon Regiment	4 squadrons
Akhtyrsk Hussars	8 squadrons
Lithuanian Ulan Regiment	8 squadrons
Horse Artillery Battery Nr 10	

	Total IV Cavalry Corps	32 squadrons 1 battery
		3,800 men, 12 guns

Reserve of the Left Wing
2nd Combined Grenadier Div — *MG Count Voronzov*
Combined Grenadier Regiments of the
2nd, 12th and 26th Divisions   2 battalions each

Total	2,100 men

The Combined Grenadier Regiments of the 7th and 24th Divisions were deployed as supports for the Jägers who were in skirmishing order along the front.

2nd Cuirassier Division — *MG Duka*
Ekaterinoslav Cuirassier Regiment
Military Order Cuirassier Regiment
Gluchov Cuirassier Regiment
Little Russia Cuirassier Regiment
Novgorod Cuirassier Regiment

Total	20 squadrons, 2,300 men

Reserve Artillery of the Left Wing
7 batteries   2,400 men, 84 guns

The Beresina Crossing

French Forces

IX Corps	***Marshal Victor, Duke of Belluno***
12th Infantry Division	***GdD Partouneaux***
1st Brigade	*GdB Camus*

- 10th Light IR — 4th Battalion
- 29th Light IR — 4 battalions

2nd Brigade — *GdB Blanmont*
- Provisional Regiment — 3 battalions
  (Made up of the 4th Battalions of the 36th, 51st and 55th Line IR)
- 125th Line IR* — 3 battalions

3rd Brigade — *GdB Billard*
- 44th Line IR — 2 battalions
- 126th Line IR* — 4 battalions
- 20th Coy, 5th Artillery Regt

* ex-Dutch regiments

26th Infantry Division — ***GdD Dändels***
1st Brigade — *GdB Damas*
- 1st Berg Line IR — 3 battalions
- 4th Berg Line IR — 3 battalions

2nd Brigade — *GdB Lingg*
- 2nd Berg Line IR — 2 battalions
- 3rd Berg Line IR — 1st Battalion

3rd (Baden) Brigade — *GdB Graf Hochberg*
- 1st Line IR — 2 battalions
- 3rd Line IR — 2 battalions
- 1st Jäger Battalion

- Berg Artillery — 1 foot and 1 horse battery
- Baden Artillery — 1 foot and 1 horse battery

28th (Polish) Infantry Division — ***GdD Girard***
Brigade structure not known. Polish authorities list the brigade commanders as:
Generals Jozef Chlopicki, Nikolai Bronikowski and *Michal Radziwill*
- 4th IR — 2 battalions
- 7th IR — 2 battalions
- 9th IR — 2 battalions
- 1 battery foot artillery

30th Light Cavalry Brigade — *Gd Fournier-Sarlovese*
- 2nd Lancers of Berg — 4 squadrons
- Hesse-Darmstadt Garde ChL Regt — 4 squadrons

31st Light Cavalry Brigade — *GdB Delaitre*
- Saxon ChL Regt Prinz Johann — 4 squadrons
- Baden Hussars — 4 squadrons

Russian Forces

Wittgenstein's Advanced Guard

There is confusion as to exactly which units which fought here. The order of battle shown in Bogdanovich is dated in the early part of the campaign and had apparently been overtaken by events by November. It is known that the Russians tinkered with the make up of their corps as the war went on. We thus show below the units shown in the action reports as well as those which were originally shown at the start of the campaign.

Commander	Maj-Gen Vlastov	
Navaginsk IR	2 battalions	
St. Petersburg Militia	7th Battalion	
26th Jäger	2 battalions	
Iamburg Dragoons	2 squadrons	
Riga Dragoons	1 squadron	
Polish Ulans	1 squadron	
Cossacks	1 squadron	
Kalmuks	1 squadron	
9th Foot Battery		12 guns
27th Position Battery		12 guns

Lt-Gen Berg's Corps	Maj-Gen Fok	
Azov IR	2 battalions	
Kaluga IR	2 battalions	
Perm IR	2 battalions	
Sievsk IR	2 battalions	
St. Petersburg Militia	2 battalions	
34th Jäger	2 battalions	
? Dragoons*	4 squadrons	
Combined Guards Cavalry Regt	3 squadrons	
Alexandria Hussars	4 squadrons	
27th Foot Battery		12 guns
5th Position Battery		12 guns

The following units were also present:
  Nizhov and Voronezh Infantry Regiments
  Depot Battalion of the Pavlov Grenadiers
  Composite cuirassier regiment made up of depot troops.

* some sources show this as being the Iamburg Dragoons but this regiment was with Wittgenstein's I Corps.

The Clash at Haynau

French Forces

16th Division — *General Count Maison*

1st Brigade — *GdB Mandeville*
  152nd Line IR — 1st, 2nd, 3rd Battalions
  154th Line IR — 1st, 2nd, 3rd Battalions

2nd Brigade — *GdB Avril*
  151st Line IR — 1st, 2nd, 3rd Battalions
  153rd Line IR — 1st, 2nd, 3rd Battalions

Artillery — 2 x FAB

Total — 3,700 men / 18 guns

3rd Light Cavalry Division — *GdD Baron Chastel*

4th Light Cavalry Brigade — *GdB Baron Vallin*
  8th Chasseurs-à-Cheval *Maj Planzeaux* — 1st & 2nd Squadrons
  9th Chasseurs-à-Cheval *Col Dukermont* — 1st & 2nd Squadrons
  25th Chasseurs-à-Cheval — 1st & 2nd Squadrons

5th Light Cavalry Brigade — *GdB Merlin*
  1st Chasseurs-à-Cheval *Col Hubert* — 1st–3rd Squadrons
  19th Chasseurs-à-Cheval *Col Vincent* — 1st–4th Squadrons
  25th Chasseurs-à-Cheval — 2 squadrons

Artillery — 1 HAB

Total — ca 2,500 men / 6 x 6-pdr guns

Prussian Forces — *General der Kavallerie von Blücher*

GM von Corswant

11th Brigade — *GM von Zieten*

Col von Dolff
  Garde du Corps — 5 squadrons
  Guard Light Cavalry Regiment — 6 squadrons

Col von Wahlen-Jürgass
  Brandenburg KürR — 4 squadrons
  East Prussian KürR — 4 squadrons
  Silesian KürR — 4 squadrons
  3 x HAB

Col von Mutius
  Neumark Dragoons — 3 squadrons

10th Brigade — *Col von Pirch I*
  1st Silesian IR — 3 battalions
  2nd Silesian IR — 3 battalions
  Silesian Grenadier Battalion — 2 companies
  Silesian Sharpshooters — 2 companies
  2 x FAB

Total — 6,400 cavalry / 8,000 infantry / 40 guns

The Clash at Liebertwolkwitz

French Forces	**Marshal Murat, King of Naples**	
II Corps	*Marshal Victor, Duke of Belluno*	
4th Division	*GdD Baron Dubreton*	
1st Brigade	*GdB Ferrière*	
24th Light IR	3 battalions	
19th Line IR	3 battalions	
2nd Brigade	*GdB Baron Brun*	
37th Line IR	3 battalions	
56th Line IR	3 battalions	
Artillery	2 x 6-pdr FABs	16 guns
5th Division	*GdD Baron Dufour*	
1st Brigade	*GdB Baron Devaux*	
26th Light IR	3 battalions	
39th Line IR	3 battalions	
2nd Brigade	*GdB ?*	
46th Line IR*	1 battalion	
72nd Line IR*	1 battalion	

These regiments had lost their artillery and been reduced from 3 to 1 battalion each at Kulm on 30 August.

Artillery		6 guns
6th Division	*GdD Baron Vial*	
1st Brigade	*GdB Marquis Valory*	
11th Light IR	3 battalions	
4th Line IR	3 battalions	
2nd Brigade	*GdB Count Bronikowski*	
2nd Line IR	3 battalions	
18th Line IR	3 battalions	
Artillery	2 x 6-pdr FABs	16 guns
Total II Corps		15,000 men, 38 guns

The II Corps light cavalry brigade (1st and 2nd Westfalian Hussars) had gone over to the allies on 23 August.

V Corps	*GdD Count Lauriston*	
10th Division	*GdD Albert*	
1st Brigade	*GdB Bachelet-Damville*	
4th Provisional Light IR	2 battalions	
139th Light IR	3 battalions	
2nd Brigade	*GdB Bertrand*	
140th Line IR	3 battalions	
141st Line IR	3 battalions	
Artillery	2 x FABs	10 guns
16th Division	*GdD Count Maison*	
1st Brigade	*GdB Mandeville*	
152nd Line IR	3 battalions	
153rd Line IR	3 battalions	
2nd Brigade	*GdB Simmer (?)*	
154th Line IR	3 battalions	
Artillery	2 x FABs	10 guns

The Clash at Liebertwolkwitz

9th Division	***GdD Rochambeau***	
1st Brigade	*GdB Lafitte*	
135th Line IR	3 battalions	
149th Line IR	3 battalions	
2nd Brigade	*GdB Baron Harlet*	
150th Line IR	3 battalions	
155th Line IR	3 battalions	
Artillery	2 x FABs	10 guns
6th Light Cavalry Brigade	*GdB Baron Dermoncourt*	
2nd Chasseurs-à-Cheval	3 squadrons	
3rd Chasseurs-à-Cheval	2 squadrons	
6th Chasseurs-à-Cheval	3 squadrons	
Artillery Reserve	2 x FABs	15 guns
	1 x HAB	8 guns
Total V Corps	12,000 infantry, 700 cavalry, 53 guns	

VIII Polish Corps	***GdD Prince Poniatowski***	
26th Division	***GdD Kamieniecki***	
1st Brigade	*GdB Sierawski*	
Vistula Legion Regiment	2 battalions	
1st Line IR	2 battalions	
16th Line IR	2 battalions	
2nd Brigade	*GdB Malachowski*	
8th Line IR	2 battalions	
15th Line IR	2 battalions	
Artillery	3 Polish batteries	14 guns
27th Division	***GdD Krasinski***	
3rd Brigade	*GdB Grabowski*	
12th Line IR	2 battalions	
14th Line IR	2 battalions	
Artillery	2 Polish batteries	8 guns
27th Light Cavalry Brigade	*GdB Uminski*	
14th Cuirassiers	2 squadrons	
Krakus	4 squadrons	
Polish Reserve Artillery	2 x FABs	16 guns
Total VIII Corps	5,400 infantry, 600 cavalry, 38 guns	

I Cavalry Corps	***GdD Marquis Latour-Maubourg***	
1st Light Cavalry Division	***GdD Berckheim***	
2nd Light Cavalry Brigade	*GdB Count Montmarie*	
16th Chasseurs-à-Cheval	2 squadrons	
1st ChL-Lanciers	2 squadrons	
2nd ChL-Lanciers	2 squadrons	
3rd Light Cavalry Brigade	*GdB Baron Picquet*	
5th ChL-Lanciers	2 squadrons	
8th ChL-Lanciers	2 squadrons	
1st Italian Chasseurs-à-Cheval	4 squadrons	
2nd Light Cavalry Division	***GdD Corbineau***	
1st Light Cavalry Brigade	*GdB Piré*	
6th Hussars	2 squadrons	
7th Hussars	2 squadrons	
8th Hussars	2 squadrons	

Orders of Battle

3rd Light Cavalry Division — *GdD Baron Chastel*
4th Light Cavalry Brigade — *GdB Baron Vallin*
  8th Chasseurs-à-Cheval — 2 squadrons
  9th Chasseurs-à-Cheval — 2 squadrons
  25th Chasseurs-à-Cheval — 2 squadrons
5th Light Cavalry Brigade — *GdB Count Merlin*
  1st Chasseurs-à-Cheval — 3 squadrons
  19th Chasseurs-à-Cheval — 4 squadrons

1st Heavy Cavalry Division — *GdD Count Bordessoulle*
1st Brigade — *GdB Sopransi*
  1st Cuirassiers — 2 squadrons
  2nd Cuirassiers — 2 squadrons
  3rd Cuirassiers — 3 squadrons
  6th Cuirassiers — 3 squadrons
2nd Brigade — *GdB Baron Bessières*
  9th Cuirassiers — 3 squadrons
  11th Cuirassiers — 3 squadrons
  12th Cuirassiers — 2 squadrons

3rd Saxon Brigade — *GM Lessing*
  Life Guard Kürassiers — 4 squadrons
  Von Zastrow Kürassiers — 4 squadrons
Artillery — 1 x French HAB — 6 x 6pdrs

    Total I Cavalry Corps — 1,350 men, 6 guns

V Cavalry Corps — *GdD Count Pajol*
9th Light Cavalry Division — *GdD Baron Subervie*
32nd Light Cavalry Brigade — *GdB Baron Klicki*
  3rd Hussars — 3 squadrons
  27th Chasseurs-à-Cheval — 4 squadrons
33rd Light Cavalry Brigade — *GdB Baron Vial*
  14th Chasseurs-à-Cheval — 3 squadrons
  26th Chasseurs-à-Cheval — 3 squadrons
    Total — 1,700 men, no artillery

5th Heavy Cavalry Division — *GdD Baron l'Héritier*
1st Brigade — *GdB Baron Queunot*
  2nd Dragoons — 3 squadrons
  6th Dragoons — 3 squadrons
  11th Dragoons — 4 squadrons
2nd Brigade — *GdB Baron Collaert*
  11th Dragoons — 4 squadrons
  13th Dragoons — 2 squadrons
  15th Dragoons — 3 squadrons
    Total — 1,700 men

6th Heavy Cavalry Division — *GdD Count Milhaud*
1st Brigade — *GdB Baron de Lamotte*
  18th Dragoons — 2 squadrons
  19th Dragoons — 2 squadrons
  20th Dragoons — 2 squadrons
2nd Brigade — *GdB Viscount de Montelegier*
  22nd Dragoons — 3 squadrons
  25th Dragoons — 3 squadrons
Artillery — 1 French HAB
    Total — 1,600 men, 6 guns

The Clash at Liebertwolkwitz

Allied Forces — **FM Fürst Karl zu Schwarzenberg**

IV Austrian Corps — *GdK Freiherr von Klenau*

1st Division — *FML Freiherr von Mohr*
1st Brigade — *GM Ritter von Paumgarten*
- Grenz-IR Wallachisch-Illyrisch Nr 13 — 2 battalions
- Grenz-IR Wallachen Nr 16 — 1st Battalion
- ChL Regt Hohenzollern Nr 2 — 4 battalions
- Hussar Regt Palatinal Nr 12 — 6 squadrons
- Artillery — 2 x HAB — 10 guns

Division of — *FML Prince Ludwig von Hohenlohe-Bartenstein*
1st Brigade — *GM von Schaeffer*
- IR Josef Colloredo Nr 57 — 2 battalions
- IR Zach Nr 15 — 3 battalions

2nd Brigade — *GM Splenyi de Milhaldy*
- IR Württemberg Nr 40 — 3 battalions
- IR Lindenau Nr 29 — 3 battalions
- Artillery — 2 x FABs — 12 guns

Division of — *FML Ritter Mayer von Heldensfeld*
1st Brigade — *GM Freiherr von Abele*
- IR Alois Liechtenstein Nr 12 — 3 battalions
- IR Koburg Nr 22 — 3 battalions

2nd Brigade — *GM de Best*
- IR Erzherzog Karl Nr 3 — 2 battalions
- IR Kerpen Nr 49 — 2 battalions
- Artillery — 2 x FAB — 12 guns

Corps Artillery Reserve
- 1 x 12-pdr battery
- 1 x 3-pdr battery — 10 guns

Total IV Corps — 24 battalions, 10 squadrons, 44 guns

Attached from the Army of Bohemia's Reserve Corps
4th Brigade — *GM Desfours*
- Kürassier Regt Kaiser Nr 1 — 6 squadrons
- ChL Regt O'Reilly Nr 3 — 6 squadrons

II Prussian Corps — *GL von Kleist*
Only part of this corps was engaged

9th Brigade — *GM von Kluex*
- 1st West Prussian IR — 3 battalions
- 6th Reserve IR — 3 battalions
- Silesian Schützen — 2 companies
- 7th Silesian Landwehr IR — 2 battalions
- Neumark Dragoons — 4 squadrons
- Artillery — 1 x 6-pdr FAB — 8 guns

10th Brigade — *GM von Pirch I*
- 2nd West Prussian IR — 3 battalions
- 7th Reserve IR — 2 battalions
- 9th Silesian Landwehr IR — 2 battalions
- 1st Silesian Landwehr Cavalry — 1 squadron
- Artillery — 1 x 6-pdr FAB — 8 guns

11th Brigade — *GM von Ziethen*
- 1st Silesian IR — 3 battalions
- 10th Reserve IR — 2 battalions
- 8th Silesian Landwehr IR — 2 battalions
- Silesian Schützen — 2 companies
- 2nd Silesian Landwehr Cavalry — 1 squadron
- Artillery — 1 x 6-pdr FAB — 8 guns

12th Brigade — *GM Prinz August von Preussen*
- 2nd Silesian IR — 3 battalions
- 11th Reserve IR — 3 battalions
- 10th Silesian IR — 2 battalions
- 1st Silesian Hussars — 2 squadrons
- 2nd Silesian Landwehr Cavalry — 1 squadron
- Artillery — 1 x FAB — 8 guns

Reserve Cavalry — *GM von Röder*
- East Prussian Kürassier Regt — 4 squadrons*
- Silesian Kürassier Regt — 4 squadrons*
- Brandenburg Kürassier Regt — 4 squadrons*
- Silesian Ulans — 4 squadrons
- 7th Silesian Landwehr Cavalry — 4 squadrons
- 8th Silesian Landwehr Cavalry — 4 squadrons
- 2nd Silesian Hussars — 2 squadrons
- Silesian National Cavalry — 2 squadrons

* with volunteer Jäger detachments

- Artillery — 2 x HABs — 16 guns

Cavalry Brigade of — *GM von Mutius*
- 7th Silesian Landwehr Cavalry — 6 squadrons
- 8th Silesian Landwehr Cavalry — 6 squadrons

Corps Artillery Reserve — *Lt Col von Braun*
- 1 x 6-pdr FAB, 1 x 12-pdr FAB
- 1 x HAB, 1 x howitzer battery — 32 guns

Total II Corps — 31 battalions, 39 squadrons, 80 guns

Russian Troops — *GdK Count Wittgenstein*

I Corps — *Lt-Gen Prince Gorshakov*

All infantry regiments had 2 battalions each

5th Division — *MG Mesentziev*
- 1st Brigade — *MG Lukov*
  - IRs Perm and Sievsk
- 2nd Brigade — *MG Vlastov*
  - IRs Kaluga, Mohilev
- 3rd Brigade — *MG ?*
  - 23rd and 24th Jägers,
  - Militia Battalion of Duchess of Oldenburg
- Artillery — 1 x FAB — 12 guns

14th Division — *MG von Helfreich*
- 1st Brigade — *MG Ljallin*
  - IRs Estonia, Tenguinsk
- 2nd Brigade — *MG Roth*
  - IRs Navaginsk, Tulsk

The Clash at Liebertwolkwitz

3rd Brigade	MG Vustov	
25th and 26th Jägers		
Artillery	1 x FAB	12 guns
Artillery Reserve	1 x FAB	12 guns
	Total I Corps	5,700 men, 36 guns

II Corps   **Lt-Gen Prince Eugen of Württemberg**
All infantry regiments had two battalions each.

3rd Division	MG Schachafskoi	
1st Brigade	Colonel Schalnitzki	
IRs Muromsk, Reval		
2nd Brigade	MG ?	
IRs Seleginsk, Czernigov		
3rd Brigade	MG ?	
20th and 21st Jägers		
Artillery	1 x FAB	12 guns
4th Division	MG Puschnitzki	
1st Brigade	MG ?	
IRs Tobolsk, Volhynia		
2nd Brigade	MG ?	
IRs Kremenchug, Minsk		
3rd Brigade	MG ?	
4th and 34th Jägers		
Artillery	1 x FAB	12 guns
Artillery Reserve	1 x FAB, 2 x HABs	36 guns
	Total II Corps	5,200 men, 60 guns

3rd Cuirassier Division*	**Lt-Gen Count Duka II**	
1st Brigade	GM Levachov	
Starodub Cuirassiers		4 squadrons
Novgorod Cuirassiers		4 squadrons
2nd Brigade	GM Gudovich	
Little Russia Cuirassiers		4 squadrons
Military Order Cuirassiers		4 squadrons

* *attached from the Russian Imperial Guard*

1st Hussar Division	**Lt-Gen Count Pahlen II**	
Brigade of	MG Rüdiger	
Grodno Hussars		6 squadrons
Olviopol Hussars		2 squadrons
Soum Hussars		6 squadrons
Brigade of	GM Lisanevich	
Lubny Hussars		4 squadrons
Chuguyev Ulans		6 squadrons
Hussars of the Guard		6 squadrons
Brigade of	Colonel Schufanov	
Serpuchov Hussars		4 squadrons
Tartar Ulans		4 squadrons
Cossacks	MG Illovaiski XII	
Pulks of Ataman, Eupatoria, Grekov VIII, Illovaiski XII, Radionov II		
	Total cavalry	2,600 men

The Battle of Möckern

French Forces

VI Corps	**Marshal Marmont Duc de Ragusa**
Chief of Staff	*GdB Richemont*
Commander of the Artillery	*GdD Foucher*
Commander of the Engineers	*Maj Constantin**

20th Division — *GdD Count Compans*
1st Brigade — *GdB Baron Pelleport*
- 32nd Light IR *Maj Cheneser* — 2nd & 3rd Battalions
- 1st Marine Artillery *Col Marechal* — 1st–5th Battalions

2nd Brigade — *GdB Baron Joubert*
- 3rd Marine Artillery *Col Bormann* — 1st, 2nd, 3rd Battalions
- 20th Prov Regt *Maj Druault* — 2 battalions
- 66th Line IR — 5th Battalion
- 122nd Line IR — 3rd Battalion
- 25th Prov Regt *Col Bochaton* — 3rd Battalion
- 47th Line IR — 3rd Battalion
- 86th Line IR — 3rd Battalion

Artillery — 2 x FABs — 12 x 6-pdr
 4 x 5.7-inch how

Total — 4,846 men, 16 guns

21st Division — *GdD Count Lagrange*
1st Brigade — *GdB Baron Jamin*
- 37th Light IR *Col Jaquet* — 1st–4th Battalions
- 4th Marine Artillery *Col Rouvroy* — 1st, 2nd, 3rd Battalions
- Regt Joseph Napoleon *Maj Dimpre* — 1st Battalion

2nd Brigade — *GdB Baron Buquet*
- 2nd Marine Artillery *Col Deschamps* — 1st–6th Battalions

Artillery — 2 x FABs — 12 x 6-pdr
 4 x 5.7-inch how

Total — 5,877 men, 16 guns

22nd Division — *GdD Baron Friederichs*
1st Brigade — *GdB Baron von Coehorn*
- 23rd Light IR *Maj Jeannin* — 3rd and 4th Battalions
- 11th Prov Regt *Col Goujon* — 2 battalions
- 1st Line IR — 4th Battalion
- 62nd Line IR — 2nd Battalion
- 13th Prov Regt *Maj Cogne* — 2 battalions
- 14th Line IR — 3rd Battalion
- 16th Line IR — 4th Battalion
- 15th Line IR *Col de Rouge* — 3rd and 4th Battalions

2nd Brigade — *GdB Choisy*
- 16th Prov Regt *Col Verbois* — 6th Battalion
- 26th Line IR — 6th Battalion
- 82nd Line IR — 6th Battalion
- 70th Line IR *Col Maury* — 3rd and 4th Battalions
- 121st Line IR *Maj Prost* — 3rd and 4th Battalions

The Battle of Möckern

Artillery	2 x FABs	12 x 6-pdr
		4 x 5.7-inch how
	Total	5,891 men
		16 guns

5th Light Cavalry Division *GdD Lorge*
12th Light Cavalry Brigade *GdB Jacquinot*
  5th Chasseurs-à-Cheval   3 squadrons
  10th Chasseurs-à-Cheval   3 squadrons
  13th Chasseurs-à-Cheval   3 squadrons
13th Light Cavalry Brigade *GdB Merlin*
  15th Chasseurs-à-Cheval   1 squadron
  21st Chasseurs-à-Cheval   1 squadron
  22nd Chasseurs-à-Cheval   2 squadrons
25th (Württ.) Lt Cav Bde *GM Graf Normann*◊
  2nd Leib-ChL *Col Prinz v. Wallerstein*   4 squadrons
  4th ChL König *Lt-Col v. Mylius*   4 squadrons
  Württemberg HAB *Lt v. Fleischmann*   6 guns
                             Total   898 men, 6 guns

◊ *Went over to the allies on 18 October with 556 men and 1 gun.*

Reserve Artillery and Park
3 x FABs   18 x 12-pdr guns, 6 x how
2 x HABs   8 x guns, 4 x how
                            Total   1,346 men, 36 guns

              Total VI Corps (on 1 October)   18,858 men
                                                          90 guns

By 16 October, this had become about 18,600 men.

* *In his memoirs Marmont states that Col Marion commanded his engineers, but Six does not confirm this*

Prussian Forces
The order of battle of Yorck's corps is taken from his biography by Droysen and differs greatly from the other sources quoted but seems to reflect the actual groupings as mentioned for 16 October by Seyfert which is highly regarded and very detailed. After the heavy losses suffered at Möckern on 16 October, Yorck's corps was slightly reorganized from the details shown here.

* = wounded on 16 October
\+ = killed on 16 October

I Corps   *Gen-Lt von Yorck*
Chief of Staff   Col von Zielinski
Adjutant *Capt Delius*
Oberquartiermeister *Lt-Col Freiherr v. Valentini*
General Staff *Maj Wilhelm v. Schack, Maj v. Klitzling, Capt v. Dedenroth,*
  Capt v. Lollhoefel, Lt v. Wussow
Adjutants *Maj Graf Brandenburg, Maj v. Diedrich, Capt v. Selasinsky,*
  Rittmeister Ferdinand v. Schack, Lt v. Below, Lt v. Röder
Kriegskommissar *v. Reiche*
Corps Medical Officer *Dr Voeltze*
Headquarters Medical Officer *Dr Hohenhorst*
Padre *Divisionsprediger Schultze*

Orders of Battle

1st Brigade   *Col von Steinmetz**
Brigade Staff *Capt v. Kaufberg+*
Adjutants *Capt v. Lützow, Lt Henkel v. Donnersmark*

1st Grenadier Brigade   *Maj Hiller von Gärtringen**
   1st East Pruss Gren Batt *Maj v. Leslie**   1 battalion
   Leib Gren Batt *Maj v. Carlowitz**   1 battalion
   Silesian Gren Batt *Maj v. Burghof*   1 battalion
   West Pruss Gren Batt *Maj v. Schon**   1 battalion

2nd Landwehr Brigade   *Col von Losthin**
   5th Silesian L'wehr IR *Maj v. Maltzahn+*   3 battalions
   Maj v. Mumm * Maj v. Seydlitz*, Maj v. Kossecky+
   13th Silesian L'wehr IR *Maj v. Gödike+*   4 battalions
   Maj v. Larisch, Maj Walter v. Cronegk, Maj v. Rekowsky+, Maj v. Martitz
   2nd Leib Hussar Regt *Maj v. Stoeffel*   4 squadrons
   2nd 6-pdr FAB *Lt Lange*   8 guns

                       Total   11 battalions, 4 squadrons
                                         9,270 men, 8 guns

2nd Brigade   *Gen-Maj Prinz Karl von Mecklenburg**
Brigade Staff *Maj v. Schütz*, Lt v. Riesenburg*
Adjutants *Maj v. Folgersberg, Capt v. Heinzmann*

1st Line Brigade   *Lt-Col von Lobenthal**
   1st East Prussian IR   3 battalions
   Maj v. Silesianeuse+, Maj v. Kurnatowsky*, Maj v. Pentzig*
   2nd East Prussian IR *Lt-Col v. Sjoeholm*   3 battalions
   Maj v. Dessauniers*, Maj v. Krauthof

2nd Landwehr Brigade
   6th Silesian Landwehr IR *Maj v. Fischer**   1 combined battalion
   M'burg-Strelitz Hussars *Lt-Col v. Warburg*   4 squadrons
   1st 6-pdr FAB *Capt Huet*   8 guns

                       Total   7 battalions, 4 squadrons
                                         7,673 men, 8 guns

7th Brigade   *Col von Horn*
Brigade Staff *Maj v. Rudolphi, Lt v. Manstein*
Adjutants *Capt Graf Kanitz, Lt v. Barfuss*, Lt v. Reibnitz**

1st Line Brigade   *Col von Zeppelin*
   Leibregiment *Col v. Zeppelin*   3 battalions
   Maj v. Örtzen, Maj v. Hagen, Maj v. Ledebur
   The Thüringian Battalion *Maj v. Linker*   1 battalion
   Gardejäger *Capt v. Bock*   ½ battalion

2nd Landwehr Brigade   *Col von Welzien*
   4th Siles. L'wehr IR *Maj Graf Herzberg*   2 battalions
   15th Siles. L'wehr IR *Col v. Wollzogen**   3 battalions
   Maj v. Sommerfeld, Maj v. Pettinger, Maj Graf Wedell
   3rd Siles. L'wehr Cav *Maj v. Falkenhausen*   2 squadrons
   Brandenburg Hussar Regt *Maj v. Sohr**   3 squadrons
   3rd 6-pdr FAB *Capt Ziegler**   8 guns

                       Total   8½ battalions, 5 squadrons
                                         8,686 men, 8 guns

The Battle of Möckern

8th Brigade   **Gen-Maj von Hühnerbein**
Brigade Staff *Capt v. Arnaud, Lt v. Unruh*
Adjutants *Lt v. Unruh, Lt v. Sellin*

1st Line Brigade   *Lt-Col von Borcke**
   Brandenburg IR   3 battalions
   Maj v. Bülow, Maj v. Othegraven*, Maj v. Krosigk+
   12th Reserve IR   3 battalions
   Maj v. Hermann, Maj v. Zeppelin, Maj v. Laurens*

2nd Landwehr Brigade
   14th Silesian Landwehr IR *Col v. Goza*   2nd and 3rd Battalions
   Maj v. Thile*, Maj v. Brixen

Cavalry
   B'burg Hussar Regt *Maj v. Knaublauch*   2 squadrons
   3rd Silesian L'wehr Cav *Maj v. Kanilowsky* 2 squadrons
   15th 6-pdr FAB *Lt Anders*   8 guns
                                 Total   8 battalions, 4 squadrons
                                                           7,447 men, 8 guns

Corps Reserve Cavalry   **Col Freiherr von Wahlen-Jürgass**
Brigade Staff *Rittmeister Freiherr v. Canitz, Lt v. Briesen*
Adjutants *Maj v. Palusdorf, Graf Reuss XIII, Graf Ingenheim*

1st Brigade   *Col Graf Henkel von Donnersmark*
   Lithuanian Drag Regt *Lt-Col v. Below*   4½ squadrons
   1st W Pruss Drag Regt *Lt-Col v. Wuthenow* 4 squadrons

2nd Brigade   *Lt-Col von Katzeler**
   Brandenburg UR *Maj v. Stutterheim*   4 squadrons
   East Prussian NCR *Maj Graf Lehndorf*   4 squadrons

3rd Brigade   *Maj von Bieberstein*
   5th Silesian L'wehr Cav *Maj v. Ozorowsky* 4 squadrons
   10th Silesian L'wehr Cav *Maj v. Sohr*   4 squadrons
   1st Neumark L'wehr Cav *Maj v. Sydow*   4 squadrons
   1st HAB *Capt v. Zinken*
   3rd HAB *Lt v. Barowsky*   16 guns
                                 Total   28½ squadrons
                                                         3,896 men, 16 guns

Corps Artillery Reserve   *Lt-Col von Schmidt*
Adjutants *Lt Erhard, Lt v. Peucker*
Artillery Staff *Maj v. Fiebig, Maj v. Rentzell, Maj v. Graumann*
Engineers *Maj Markoff, Lt v. Huelsen, Lt v. Poser*
   1st 12-pdr FAB *Lt Witte*
   2nd 12-pdr FAB *Lt Simon*
   12th 6-pdr FAB *Lt Bully*
   24th 6-pdr FAB *Lt Varenkampf*
   1st 3-pdr FAB *Lt v. Oppen*
   3rd HAB *Lt Fischer*
   12th HAB *Rittmeister v. Pfeil*
                                   Total   1,248 men, 56 guns
                   Total I Corps   34½ battalions, 45½ squadrons
                                                           38,220 men
                                                       13 batteries, 104 guns

All totals include detached troops. Actual strengths at Leipzig were therefore much lower than shown here.

Allied Cavalry Raids of 1813

Army of Bohemia
Streifkorps of   *Lt-Gen von Thielemann*
Lieutenant-Col Freiherr von Gasser

ChL Regt Hohenzollern Nr 2	2 squadrons	210 men
ChL Regt Klenau Nr 5	1 squadron	140 men
Hussar Regt Kienmayer Nr 8	1 squadron	100 men

General-Major Prinz Biron von Kurland

Silesian Hussar Regt Nr 2	2½ squadrons	200 men
2nd Silesian National Cavalry Regt	2 squadrons	250 men
Neumark Dragoon Regiment	Freiwillige-Jäger squadron	?

Cossacks *Col Count Orlov-Denisov*

Pulks of Gorin II and Jagodin II		600 men
Artillery	2 x Austrian & 2 x Cossack HA how	
	Total	9½ squadrons, 2 pulks 1,500 + ? men, 4 guns

Streifkorps of   *Col Count von Mensdorff-Poully*

Hussar Regt EH Ferdinand Nr 3	2 squadrons	260 men
Hesse-Homburg Hussar Regt Nr 4	1 squadron	130 men
Don Cossack pulks of Gorin I and Illovaiski X		400 men in all
	Total	3 sqns, 2 pulks, 790 men

Streifkorps of   *Maj-Gen Illovaiski XII*
Don Cossack pulks of Grekov I, Grekov VIII and Illovaiski XII

Total	850 men

Army of Silesia
Streifkorps of   *Major von Boltenstern*

Guard-Jäger Battalion	1 coy + Freiwillige-Jäger det'ment
Neumärk Landwehr Cavalry	½ squadron
1st Ukrainian Cossack Regiment	½ squadron
Total	250 infantry, 80 cavalry.

Streifkorps of   *Captain Count Puckler*

Brandenburg Hussar Regiment	Freiwillige-Jäger detachment
3rd Ukranian Cossack Regiment	½ squadron
Total	100 men.

Streifkorps of   *Major von Colomb*
Two combined cavalry squadrons from the II Prussian Corps, including the Freiwillige-Jäger of the Brandenburg Hussars.

Total	170 men.

Royal Prussian Freikorps   *Col von Lützow*
This corps was raised by royal order of 18 February 1813. On 4 June it had three battalions and five squadrons; during the armistice eight guns were added and the final form was:

Infantry	3 battalions + Jäger detachment
Tyrolean Jäger	1 company
Cavalry regiment	5 squadrons, inc. 3 of hussars, 1 of mounted rifles and 1 of Ulans
Foot artillery	1 battery
Horse artillery	1 battery

Allied Cavalry Raids of 1813

The corps fought at the Göhrde on 12 May, Kitzen on 17 June as described in the text, Lauenburg 17–19 August, Gadebusch on 26 August, Moellen on 5 September, Zarrentin on 18 September, Bremen on 13 and 14 October, on the Stecknitz on 8 November, blockade of Glückstadt 21–25 December, and the siege of Jülich fortress February–4 May 1814. On 19 January 1814 the infantry were used to form Lützow's Line IR (later the 25th Regiment). The cavalry went to the 6th Ulans and the 9th Hussars; the artillery to the Silesian Artillery Brigade.

Freikorps of   *Major von Hellwig*

On 23 April 1813 Major von Hellwig of the 2nd Silesian Hussars was ordered to raise a Freikorps. As a basis for the corps he was allowed to use the 3rd and 4th Squadrons of his own regiment. On 25 May he was joined by Captain von Barthel with 60 infantrymen. By the end of November the corps had grown to three squadrons of hussars, one of mounted rifles, three companies of light infantry and a detachment of Jägers. On 18 April 1814 the two squadrons of Silesian Hussars went back to their regiment. They were replaced in the corps by two squadrons of hussars raised in Hamburg by Major von Schill in 1813. Hellwig's corps was disbanded in March 1815; the infantry went into the newly-formed 27th Infantry Regiment, the cavalry to the new 7th Ulans.

Army of the North

Streifkorps of   *Lt-Col Davidov*

Denis Davidov, Lt-Col of the Akhtyrsk Hussars commanded a small, mixed Streifkorps of about 100 hussars of his own regiment and 400 Cossacks in Voronzov's Avantgarde of Winzingerode's Russian corps. It operated under the command of Maj-Gen Lanskoi of the 2nd Hussar Division.

Streifkorps of   *Gen-Maj Tettenborn*
1 x Cossack pulk   approx 800 men*
                                                                  *in October

Streifkorps of   *Gen Czernichev*

This force was put together fairly late in the campaign, when the successes of the other raiding groups had been proved.

Cavalry Division   *Maj-Gen Magnus Pahlen*
   Riga Dragoons
   Finland Dragoons
   Isum Hussars
   Cossack pulks of Barabanchikov II, Grekov IX, Illovaiski IV and Lochilin I
                                             Total   2,300 men and 6 light guns

Cossack Raiding Corps   *General of Cavalry Count Platov*
   Pulks of Grekov VIII, Illovaiski X,
   Kaisarov, three other pulks
   Artillery   1 x Don Cossack HAB   12 guns

Brigade of   *Maj-Gen Seslavin*
   Soum Hussars   2 squadrons
   4 Cossack pulks
   Artillery   2 guns

The Battle of Fère-Champenoise

Many details of the French forces which fought here, which divisions and regiments were involved and who were the commanders of the various formations, remain very unclear, particularly with regard to the cavalry formations. For example, various sources show divisions commanded by generals who do not appear in Six's biographical dictionary. Other battle reports show regiments which theoretically were not there, and other regiments supposedly not present are listed by Martinien as having officer casualties in the battle.

French Forces   **Marshal Mortier**

The Young Guard   *Maj-Colonel Count Michel*
Fusilier-Grenadiers	1 battalion
Fusilier-Chasseurs	2 battalions
Velites de Florence	½ battalion
Velites de Turin	2 companies
Flanquers-Grenadiers	1 battalion
Flanquers-Chasseurs	2 battalions
1st, 5th, 6th, 7th, 8th, 14th Tirailleurs	
1st, 2nd, 3rd, 4th, 5th 7th, 8th, 14th Voltigeurs	
Artillery	not known
Total	3878 men

VI Corps   *Marshal Marmont*

1st Division   *GdD Count Ricard*
The brigading of these units is not known
2nd, 4th, 6th, 9th, 16th Light IR	1 or 2 companies each
22nd, 40th, 50th, 69th, 136th, 138th, 142nd, 144th, 145th Line IR	1–4 companies each
Total	2,917 men, 7 guns

2nd Division   *GdD Count Lagrange*
The brigading of these units is not known
1st, 2nd, 3rd, 4th Naval IR	1 battalion each
62nd and 132nd Line IR	3 companies each
28th, 43rd, 48th, 54th, 65th, 70th, 112th, 149th Line IR	1 battalion each
Total	4,868 men, ? guns

1st Light Cavalry Division   **GdD Count Doumerc**
Brigade of   *Count Doumerc*
1st Provisional Hussars	4 squadrons
2nd Chasseurs-à-Cheval	4 squadrons
3rd Chasseurs-à-Cheval	4 squadrons
4th Chasseurs-à-Cheval	4 squadrons
	1,900 men

Brigade of   *GdB Picquet*
1st Gardes d'Honneur	4 squadrons
10th Hussars	4 squadrons
	915 men

Artillery
Horse Artillery	191 men, ? guns
Foot Artillery	539 men, ? guns

The Battle of Fère-Champenoise

I Cavalry Corps	**GdD Auguste-Danielle Count Belliard***
	* In place of Grouchy, wounded

1st Heavy Cavalry Division — **GdD Count Bordessoulle**
1st Brigade — **GdB Thiry**
  2nd, 3rd, 6th, 9th, 11th, 12th Cuirassiers   2 squadrons each

2nd Brigade — **GdB Count Laville**
  4th, 7th, 14th Cuirassiers   2 squadrons each
  7th, 23rd, 28th, 30th Dragoons   2 squadrons each

2nd Heavy Cavalry Division — **GdD Baron Berckheim**
Under Belliard from 17 March. It is not known which regiments were in this division in this action nor how many squadrons of each regiment listed were present at the battle.

1st Light Cavalry Division — **GdD Count Merlin**
1st Brigade — **GdB Wathiez**
  6th, 7th, 8th Hussars
  1st, 3rd, 5th, 7th, 8th ChL-Lanciers

2nd Brigade — **GdB Baron Guyon**
  1st 2nd, 3rd, 6th, 8th,
  9th, 16th, 25th Chasseurs-à-Cheval   1 or 2 squadrons each

6th Dragoon Division — **GdD Roussel d'Hurbal**
From Kellermann's VI Cavalry Corps; under Belliard from 17 March.

11th Light Cavalry Brigade — **GdB Count Sparre**
  5th Dragoons   4 squadrons
  12th Dragoons   4 squadrons

12th Light Cavalry Brigade — **GdB Rigaud (?)**
  21st Dragoons   4 squadrons
  26th Dragoons   4 squadrons

The following regiments are listed as having officer casualties at Fère-Champenoise, but were apparently not part of I Cavalry Corps: 5th, 6th, 21st, 25th, 26th Dragoons, 23rd Chasseurs-à-Cheval.

French total — generally accepted as 12,700 infantry, 4,000 cavalry, 84 guns. Other sources give 25,000 men including '700 cavalry'

Other French Forces

The following Garde-National units were organised into two small divisions under Generals Amey and Pacthod. They were not involved in the action at Fère-Champenoise, but were destroyed by the corps to which the allied cavalry belonged that same day, just off the field to the north.

Garde-National Units

Cherbourg	2nd Battalion
Eure-et-Loire	Elite-Battalion
Indre-et-Loire	Elite-Battalion
Rochefort	Elite-Battalion
Seine	1st Elite-Battalion
Seine-et-Marne	Elite-Battalion
Total	4,300 men, ? guns

Allied Forces — Crown Prince Wilhelm of Württemberg

IV (Württemberg) Corps — Gen-Lt Baron von Koch

Brigade of — *General-Major von Walsleben;*
Prince Adam of Württemberg
  2nd Jäger-zu-Pferde Herzog Louis   4 squadrons
  3rd Dragoons Kronprinz   4 squadrons
Artillery   1 x 6-pdr HAB

Brigade of — *General-Major von Jett*
  4th Jäger-zu-Pferde Prinz Adam   4 squadrons
  5th Jäger-zu-Pferde   4 squadrons
Artillery   1 x 6-pdr HAB

Russian Imperial Guard — *Gen of Cavalry Grand Duke Constantine*
Lt-Gen Count Orlov-Denisov

Brigade of — *MG Chalikov*
  Life Guard Dragoons   4 squadrons
  Life Guard Hussars   4 squadrons
  Life Guard Ulans   4 squadrons
  Life Guard Don Cossacks   4 squadrons
  Life Guard Black Sea Cossacks   4 squadrons

1st Cuirassier Division — *Lt-Gen Depreradovich*

Brigade of — *MG Arseniev*
  Chevalier Guards   4 squadrons
  Horse Guards   4 squadrons

Brigade of — *MG Rosen*
  Life Guard Cuirassiers   4 squadrons
  Cuirassiers of Her Majesty   4 squadrons

2nd Cuirassier Division — *MG Kretov*

Brigade of — *MG Leontiev*
  Pskov Cuirassier Regiment   5 squadrons
  Glochov Cuirassier Regiment   5 squadrons

Brigade of — *MG Karachev*
  Ekaterinoslav Cuirassier Regiment   4 squadrons
  Astrakhan Cuirassier Regiment   4 squadrons

Cossack Corps — *Gen of Cavalry Count Platov*
  Pulks of Grekov VIII, Illovaiski X,
  Kaisarov, three other pulks
Artillery   1 x Don Cossack HAB   12 guns

Brigade of — *MG Seslavin*
  Soum Hussars   2 squadrons
  Cossacks   4 pulks
Artillery   2 guns

The Battle of Fère-Champenoise

Cavalry of VI Corps	**Lt-Gen Count Pahlen III**	
Brigade of	MG von Rüdiger	
Grodno Hussars	6 squadrons	
Soum Hussars	4 squadrons	
Brigade of	MG Docturov	
Lubny Hussars	4 squadrons	
Olviopol Hussars	4 squadrons	
Brigade of	MG Lisanevich	
Chuguyev Ulans	6 squadrons	
Brigade of	MG Illovaiski XII	
Cossacks	5 pulks	
Artillery	1 x HAB	12 guns
Austrian Troops	**FML Count von Nostitz**	
1st Kürassier Division	**FML Count von Klebelsberg**	
Brigade of	GM Count von Auersperg	
Kürassiers Constantine Nr 8	4 squadrons	
Kürassiers Sommariva Nr 5	4 squadrons	
Brigade of	GM Count von Delfours	
Kürassiers Kaiser Nr 1	4 squadrons	
Kürassiers Liechtenstein Nr 6	4 squadrons	
Brigade of	GM von Trenck	
Chevauxlégers Rosenberg Nr 6	6 squadrons	
Hussars EH Ferdinand Nr 6	6 squadrons	
Grenz-IR Warasdiner St. Georger Nr 6	1 battalion	
Artillery	1 x HAB	
	Allied total	12,000 men, 48 guns

The Battle of Waterloo

French I Corps — *Lt-Gen Count d'Erlon*

1st Division — *Lt-Gen Allix*
54th, 55th, 28th, 105th Line IR — 8 battalions

2nd Division — *Lt-Gen Baron Donzelot*
13th Light IR 17th, 19th, 51st Line IR — 9 battalions

3rd Division — *Lt-Gen Baron Marcognet*
21st, 46th, 25th, 45th Line IR — 8 battalions

4th Division — *Lt-Gen Count Durutte*
8th, 29th, 85th, 95th Line IR — 8 battalions

1st Light Cavalry Division — *Lt-Gen Baron Jacquinot*
3rd Chasseurs, 7th Hussars — 6 squadrons
3rd and 4th ChL-Lanciers — 5 squadrons

Artillery
5 x FABs, 1 HAB — 46 guns

Total — 16,200 infantry, 1,400 cavalry, 1,066 artillerymen, 46 guns

Officer Losses of I Corps on 18 June 1815

	Killed	Mortally wounded	Wounded	Remarks
8th Line IR	1	–	19	
13th Light IR	5	1	22	
17th Line IR	4	2	15	
19th Line IR	7	2	13	
25th Line IR	–	1	30	
28th Line IR	5	1	11	
29th Line IR	1	1	8	
45th Line IR	3	–	28	Eagle taken by Sgt Ewart, Scots Greys
46th Line IR	1	2	21	
51st Line IR	6	2	11	
54th Line IR	–	6	14	
55th Line IR	1	–	14	
85th Line IR	3	2	17	
95th Line IR	1	1	17	
105th Line IR	11	1	20	Eagle taken by Capt Clark, R. Dragoons
3rd Chasseurs	1	–	10	
7th Hussars	–	–	9	
3rd ChL Regt	1	1	6	
4th ChL Regt	1	2	6	

The Battle of Waterloo

Losses of Allied Cavalry

	Strength	Killed	Wounded
1st Cavalry Brigade	*Maj-Gen Lord Edward Somerset*		
1st Life Guards	228	18	43
2nd Life Guards	231	17	41
Royal Horse Guards (Blues)	237	17	60
1st (King's) Dragoon Guards	394	43	104
2nd or Union Cavalry Brigade	*Maj-Gen Sir William Ponsonby*		
1st (Royal) Dragoons	394	89	97
2nd Dragoons (Scots Greys)	391	102	97
6th (Inniskilling) Dragoons	396	83	116

French Cavalry Officer Losses in the Last Great Failure

	Killed	Mortally wounded	Wounded
Guard Heavy Cavalry	*Lt-Gen Count Guyot*		
Grenadiers-à-Cheval	2	–	17
Guard Light Cavalry	*Lt-Gen Lefebvre-Desnouëttes*		
Chasseurs-à-Cheval	5	1	14
1st ChL Regt	–	–	–
2nd ChL Regt	1	–	9
1st Light Cavalry Division	*Lt-Gen Baron Jacquinot*		
See above			
III Cavalry Corps	*Lt-Gen Kellermann*		
11th Cavalry Division	*Lt-Gen l'Héritier*		
2nd Dragoons	5	1	11
7th Dragoons	–	2	15
8th Cuirassiers	–	–	4
11th Cuirassiers	1	–	15
12th Cavalry Division	*Lt-Gen Roussel d'Hurbal*		
1st Carabiniers	7	–	14
2nd Carabiniers	1	1	9
2nd Cuirassiers	–	1	14
3rd Cuirassiers	1	1	11
IV Cavalry Corps	*Lt-Gen Count Milhaud*		
13th Cavalry Division	*Lt-Gen Watier*		
1st Cuirassiers	4	–	13
4th Cuirassiers	3	1	10
7th Cuirassiers	2	1	11
12th Cuirassiers	2	–	12
14th Cavalry Division	*Lt-Gen Baron Delort*		
5th Cuirassiers	2	–	12
6th Cuirassiers	–	–	16
9th Cuirassiers	1	–	11
10th Cuirassiers	–	1	12

Index of Persons

Abele, Freiherr von, Aus. GM 285
Abercrombie, gen. 78, 85, 87, 89, 267
Albert, Joseph-Jean-Baptiste, gen. 282
Alexander I, Czar 45, 51, 52, 62, 158, 190, 215
Alexopol, Rus. gen. 141
Allix de Vaux, Jacques-Alexandre-François, gen. 204, 205, 229, 230, 231, 298
Almeras, Louis, gen. 273
Alten, Sir Karl, gen. 78, 87, 267
Altmann, corp. of Aus. dragoons 42
Amey, François-Pierre-Joseph, gen. 219, 220, 261, 295
Andreossy, Victor-Antoine, gen. 249
Anna Leopoldovna, Rus. princess 16
Anson, George, gen. 116, 118, 119, 271
d'Antiste, Dufforc, gen. 95,
Arbuthnot, ADC 85
Arnold, Prus. hussar trooper 132
Arrighi, Jean-Toussaint de Casanova, gen. 199, 200
Arseniev, Rus. gen. 74, 264, 296
Auer, von, gen. 266
Auersperg, Graf von, GM 297
Augereau, Charles-Pierre-François, marshal 63, 66, 67, 70, 71, 72, 74, 167, 174, 176, 213, 261
d'Aumont, Dion, sergeant 95
Autie, col. 248
Avril, gen. 281

Bachelet-Damville, Louis-Alexandre, gen. 282
Bachelu, Gilbert-Désiré-Joseph, gen. 229, 231
Bachmetiev I, Rus. gen. 141, 275
Bachmetiev II, Rus. gen. 141, 275
Bagration, Piotr, Rus. gen., prince 51, 56, 57, 125, 255, 258, 277
Ballasteros, Fransisco, gen. 78, 85, 268
Barail, gen. du 238
Barbanegre, Joseph, col. 249
Barbier, Jean-François-Thérèse, col. 254
Barclay de Tolly, Mikhail Andreovich, FM prince 65, 66, 74, 129, 135, 136, 143, 144, 158, 224, 264
Barrie, Jean-Léonard, col. 248
Barthélemi, Nicolas-Martin, col. 250
Bashmakov, Rus. officer, 143
Bastineller, Westphalian gen. 204
Bataglia, col. 272
Baurth, gen. 274
Beaumont, Marc-Antoine Bonin de La Boninière, gen. comte 251, 254
Becker, col. 251
Behm, Werner, Meck'burg hussar 188
Beker, Nicolas-Léonard Bagert, col. 252
Belfort, Jacques Belfort Renard, col. 253
Bellegarde, F. Aus. FML 35, 38, 43, 64, 208
Belliard, Auguste-Daniel, gen. 216, 253, 295
Benkendorf, col. von 197
Bennigsen, Levin August Theophil Russian gen., count 51, 62, 63, 65, 66, 67, 70, 72, 73, 74, 102, 263
Benzien, Gefreiter 189
Berckheim, Sigismond-Frédéric, gen. 167, 171, 274, 283, 295
Beresford, William Carr, marshal viscount 77, 78, 80, 81, 82, 84, 85, 87, 89, 90, 91, 92, 267

Berg 1, Grigori, Rus. gen. 257
Bernadotte, Jean-Baptiste, Fr. marshal, later appointed King of Sweden 56, 57, 59, 62, 64, 65, 74, 165, 166, 183, 185, 187, 210, 248
Bernkopf, Aus. infantry captain 28
Berthier, Louis-Alexandre, marshal 26, 28, 33, 35, 43, 44, 47, 48, 52, 139, 243, 247
Berthier, Victor-Léopold, gen. 248
Bertrand, Edme-Victor, gen. 282
Bertrand, Henri-Gatien, gen. 223
Bessières, Bertrand, col. 251, 284
Bessières, Jean-Baptiste, marshal 40, 42, 72, 97, 163, 247, 260
Best, de, Aus. GM 285
Bianchi, Aus. gen. 224
Bieberstein, major von 291
Bigarre, Auguste-Julien, col. 250
Bijlandt, gen. van 226, 227, 228
Billard, Pierre-Joseph, gen. 279
Binot, Louis-François, gen. 74
Bismark, major von 189
Blake, Joachim, gen., 78, 80, 81, 87, 91, 92, 268
Blakeney, col. 89
Blanmont, Marie-Pierre-Isidore, de, gen. 279
Blücher, Gebhard Lebrecht, FM von, 62, 106, 158, 159, 161, 162, 165, 166, 183, 185, 190, 198, 210, 211, 212, 213, 214, 215, 223, 224, 225, 281
Bochaton, col. 288
Bock, Eberhard von, gen. 116, 119, 120, 271
Boltenstern, von, major 292
Bonnard, Ennemond, gen. 63
Bonnet d'Honnières, Joseph-Alphonse-Hyacinthe-Alexandre, gen. 74, 114, 252
Borcke, von, lt-col. 291
Bordesoulle, Étienne-Tardif comte de Pommeroux, gen. 284, 295
Borman, col. 288
Borosdin II, Rus. gen. 277, 278
Boudet, Jean, gen. 30, 32, 41, 244
Bourcier, François-Antoine-Louis, gen. 52, 53, 250, 254
Bourdon, col. 253
Bourmont, Louis-Auguste-Victor de Ghaisnes gen. comte de 212, 238
Boussart, André-Joseph, gen. 253, 262
Bouvier des Eclaz, Joseph, gen. 80, 270
Boye, Charles-Joseph, gen. 251, 262
Brack, Fortune, retired captain 233
Braun, von lt-col. 286
Bredow, von, major 170, 181
Briche, André-Louis-Elisabeth-Marie, gen. 270
Brie, de, Aus. gen. 42, 245
Bron, André-François, de Bailly, gen. 80, 270
Bronikowski, Nikolai, Polish gen. 279, 282
Brooke, major, 94
Brouard, Étienne, gen. 260
Broussier, Jean-Baptiste, gen. comte de 138, 273
Brune, Guillaume-Marie-Anne, marshal 224
Brun, gen. 282
Bruyères, Pierre-Joseph, gen. 70, 71, 163, 263
Buchholz, Prus. lt-col. 162

Bülow, Friedrich Wilhelm, gen. 183, 213, 214
Buquet, Charles-Joseph, gen. 288
Burthe, André, col. 254
Buxhowden, Rus. gen. 51, 53, 58, 59, 62, 255

Cadoudal, Georges 45
Caffarelli du Falga, Marie-François-Auguste, gen. 56, 57, 249, 253
Cambronne, Pierre-Jacques-Étienne, gen. vicomte 223
Camus, Louis, gen. 279
Candras, Jacques-Lazare de Savettier de, gen. 251, 260
Caramely, Aus. gen. 257
Carl, Prince, Aus. Grand Duke 19
Carneville, Aus. gen. 255
Carra St. Cyr, Claude, gen. 196, 197, 243
Carrie de Boissy, Jean-Augustin, col. 253
Castaños, Francisco Xavier, gen. 78
Castex, col. 252
Caulaincourt, Auguste-Jean-Gabriel de, gen. 133, 135, 137, 138, 140, 250, 274
Cavaignac, Jacques-Marie, col. 253
Chabran, Joseph, gen. 27, 31, 32
Chalikov, Rus. gen. 272, 296
Chambarlhac, Jacques-Antoine de Laubespin, gen. 33, 36, 243
Chambray, marquis de 138
Champeaux, Pierre-Clement, gen. 39, 41, 43, 244
Chanetot, col. 248
Chaplitz, Rus. gen. 151, 258, 264
Chariton, Cossack commander 272
Chastel, Louis-Pierre-Aimé, gen. 128, 132, 143, 144, 159, 161, 274, 281, 284
Chemineau, Jean, gen. 118
Chernich, Rus. gen. 272
Chichagov, Pavel, Rus. admiral, 147, 150, 151, 155, 156
Chlopicki, Jozef, Polish gen. 279
Choisy, Jacques-Robert, gen. 288
Claparède, Michel-Marie, gen. 63, 134, 145, 252
Clark, captain 230, 298
Clausewitz, Carl von, gen. 128
Clauzel, Bertrand, gen. 114, 224
Cleeve, RA officer, 82, 84, 95
Clement, col. 252
Clouet, gen. 238
Cochois, Antoine-Christophe, col. 253
Coehord, Adjt Cmdt, 252
Coehorn, Louis-Jacques, gen. 288
Coignet, Jean-Roche, author 36
Colbert-Chabanais, Pierre-David, gen. 111, 239, 240, 261
Colborne, gen. 78, 82, 83, 84, 85, 92, 93, 94, 267
Cole, Sir Galbraith, gen. 78, 81, 85, 88, 90, 92, 267
Collaert, Jean-Antoine, gen. 284
Collins, gen. 78, 87, 268
Colomb, Major von 196, 292
Compans, Jean-Dominique, gen. 191, 253, 297
Conroux, Nicolas-François, col. 252
Constantine, Grand Duke 56, 58, 218, 219, 255, 258, 277, 296
Corbineau, Claude-Louis-Esprit-Juvenal, gen. 74, 254, 283

Index of Persons

Corswant, gen. 281
Cotton, Sir Stapleton, gen. 120
Coutin, Joseph-Christophe, col. 247
Crovi, col. 272
Curial, Philippe-Jean-Baptiste-François, col. 252
Curto, Jean-Baptiste-Theodore, gen. 115, 116, 271
Czeckyra, Prus. trooper 179, 181
Czepelev, Dmitri, Rus. gen. 257
Czernichev, Rus. gen. 20, 197, 199, 200, 203, 204, 205, 293

Dändels, Herman Wilhelm, gen. 153, 155, 279
Dahlmann, Nicolas, gen. 71, 74, 260
Dalwigk, von, Hessian col. 155
Damas, François-Étienne, gen. 154, 279
Damm, Rittmeister 189
Dampierre, Achille, 35, 38
Daru, Intendant-General 247
Davidov, Denis, 70, 71, 72, 197, 293
Davout, Louis-Nicolas, marshal 12, 20, 52, 53, 59, 61, 63, 65, 67, 70, 73, 205, 102, 197, 249, 260
De Billy, Jean-Louis, gen. 252, 260
Debroc, Armand-Louis, col. 252
Decaen, Charles-Mathieu-Isidore, gen. 224
Deschamps, col. 288
Decken, Gustavus von, 117, 119, 120
Defrance, Jean-Marie-Antoine, gen. 135, 138, 141, 144, 274
De Grey, col. 267
Delaitre, Antoine-Charles-Bernard, gen. 146, 279
Delfours, Graf von, GM 297
Delort, Jacques-Antoine-Adrien, gen. 299
Delzons, Alexis-Joseph, gen. 126, 145, 163, 272
Demont, Joseph-Laurent, gen. 252
Denisov VII, Cossack commander 272
Depreradovich I, Rus. gen. 259, 296
Depreradovich II, Rus. gen. 258
Deriaguine, Rus. dragoon 66
Dermoncourt, Paul-Ferdinand-Stanislas, gen. 283
Desjardin, Jacques-Jardin, gen. 74, 261
Desailly, Jean-Charles, col. 249
Desaix, Louis-Charles-Antoine, gen. 24, 25, 30, 31, 32, 36, 37, 40, 41, 42, 43, 44, 45, 98, 243
Desfours, Aus. GM, 171, 285
Detmer, surgeon, 118
Devallant, major 206
Devaux, Marie-Jean-Baptiste-Urbain, Adjt-Cmdt chevalier 252
Devaux, Pierre, gen. 282
Diebitsch, von, gen. 157
Dierecke, von, gen. 265
Digeon, Alexandre-Elisabeth-Michel, col. vicomte 251
Docturov, Rus. gen. 297
Docturov, Dmitri, Rus. gen. 59, 64, 67, 75, 134, 173, 256, 264, 276
Döring, gen. von 199
Dolgoruki, Gen Adjt 255
Dolgoruki, Rus. gen. 258
Dolff, von, Prus. col. 159, 162, 281
Dommanget, Jean-Baptiste, gen. 134, 144, 274
Donzelot, François-Xavier, gen. 298
Doriel, major 273
Doumerc, Jean-Pierre, col. 253, 294

Drouet d'Erlon, Jean-Baptiste, gen. 59, 222, 225, 227, 229, 248, 298
Drouot, Antoine, gen. 223
Dubreton, Jean-Louis, gen. 282

Dufour, François-Marie, gen. 282
Duka, Rus. gen. 172, 278
Duka II, Rus. gen. 287
Dukermont, col. 281
Dumoulin, Charles, gen. 248
Dumoustier, Pierre, col. 252
Dupas, Peirre-Louis, gen. 248
Dupont, Pierre de Letang, gen. 243
Duroc, Géraud-Christophe-Michel, gen. 163, 247, 252
Durosnel, Antoine-Jean-Auguste-Henri, col. 254, 262
Durutte, Pierre-François-Joseph, comte, gen. 298
Duvigneau, Bernard-Étienne-Marie, gen. 44, 244

Eblé, Jean-Baptiste, gen. 150, 151, 248
Edighofen, Jean-Georges, col. 251
Eiler, Rus. col. 277
Elsnitz, Anton von, Aus. gen. 245
Engelhardt, Rus. gen. 258
Enghien, duke of, 46
Eppler, Georges-Henri, gen. 252
l'Eritage, gen. 274
España, Carlos de, gen. 78, 87, 269
Essen II, Rus. gen. 56, 62, 70, 257
Essen III, Rus. gen. 265
l'Estocq, Anton Wilhelm von, gen. 62, 63, 64, 65, 67, 73, 265
Eugène de Beauharnais, viceroy of Italy 107, 123, 124, 125, 126, 128, 132, 133, 134, 135, 138, 144, 208, 211, 272
Ewart, sergeant 230, 298

Farine de Creux, Pierre-Joseph, gen. 231
Fauconnet, Jean-Louis-François, gen. 253, 254, 255
Ferdinand VII, King of Spain 210
Ferdinand, Archduke 50
Ferey, Claude-François, gen. 114, 115, 250, 261
Fereye, col. 250
Ferrière, Jacques-Martin-Madelaine, gen. 273, 282
Fery, Jean-Baptiste-Michel, gen. 251
Figner, Rus. hero 175
Fiteau, Edme-Nicolas, col. 253
Fok, Rus. gen. 280
Folgersberg, major von, 176, 178, 179
Fontaine, François-Xavier-Octavie, Adjt-Cmdt 254
Foucher de Cariel, Louis-François, col. 252, 288
Fouler, Albert-Louis-Emmanuel, col. 254
Fournier de Loysonville Daultane, gen. 249
Fournier-Sarlovese, François, gen. 146, 153, 154, 155, 185, 200, 201, 279
Foy, Maximilian-Sebastien, gen. 115, 117, 118, 121, 271
Franceschi-Delonne, Jean-Baptiste, col. 251
Franz I Emperor of Austria 108
Frere, Bernard-Georges-François, gen. 248
Friant, Louis, gen. 52, 73, 249, 260
Friederichs, Jean-Parfait, gen. 288
Friedrich, King of Württemberg 201
Friedrich Wilhelm III, King of Prussia 61, 108, 157
Frierenberger, Aus. major 57
Frimont, Aus. gen. 33, 35, 36, 40, 224
Fuemetty, von, lt. 118

Gaffron, Hermann von, Prus. officer 176
Galitzin, prince, Rus. gen. 63, 70, 71, 264, 277
Galpoy, trooper Prus. hussars 132
Gardanne, Gaspard-Amedée, gen. 32, 33, 35, 37, 38, 243, 261

Gasser, lt-col. von, 292
Gautier, Jean-Joseph, gen. 260
Gauthrin, Pierre-Edme, gen. 274
Gay, col. 249
Gazan, Honoré-Théodore-Maxime, gen. 81, 85, 90, 252, 269
Gency, Claude, gen. 243
Genty, Berg col. 154
Gérard, Maurice-Étienne, gen., 44, 134, 135, 138, 144, 213, 274
Gersdorff, Rus. gen. 75
Girard, Jean-Baptiste, gen. 81, 82, 85, 87, 89, 92, 93, 153, 154, 269, 279
Gneisenau, August Wilhelm Anton, Graf Neithardt von, gen. 212, 224
Godinot, Deo-Gratias-Nicolas, gen. 80, 87, 269
Göcking, lt-col. von 198
Gorshakov, Prince, Rus. gen. 277, 286
Gottesheim, Aus. GM, 30, 35, 36, 38, 39, 246
Gough, sergeant 84
Gourgaud, major 167
Gouvion St. Cyr, Laurent, marquis de, marshal 147, 167
Goza, von, col. 291
Grabowski, Polish gen. 283
Grandeau, Louis-Joseph, gen. 260
Grekov, Rus. Cossack, 176
Grekov VIII, Cossack, 296
Grekov XVIII, Cossack, 272, 287
Grouchy, Emmanuel, marshal, marquis de, 67, 71, 132, 144, 226, 233, 237, 239, 262, 274
Grüner, gen. 163
Gudin, Étienne, gen. 249, 260
Gudovich, Rus. GM, 287
Guenand, Louis, gen. 243
Guiton, Adrian-François-Marie, col. 254
Guylai, Aus. gen. 211
Guyon, Claude-Raymond, gen. 272, 295
Guyot, Claude-Étienne, gen. 233, 254, 299

Haddick Karl, von Futak, Aus. FML 28, 35, 244, 245
Hamilton, gen. 78, 81, 87, 268
Hammerstein, gen. von 196
Hammes, Baden lt. 152
Hanicque, Antoine-Alexandre, gen. 253
Hardegg, Aus. major, 37, 38, 43, 246
Hardinge, Henry, col. 88
Harlet, Louis, gen. 283
Harvey, gen. 82, 88, 89, 90, 268
d'Hautpoul, Jean-Joseph-Ange, gen. 67, 71, 74, 254, 262
Hayelle, col. 253
Haxo, François-Nicolas, gen. 44
Heldensfeld, FML Mayer von 285
Helfreich, Rus. gen. 172, 286
Hellwig, major von 198, 293
Henkel von Donnersmark, col. 291
Herbin-Dassaux, Jean-Baptiste, gen. 243
l'Héritier, Samuel-François, gen. 167, 168, 170, 171, 284, 299
Hessen-Philippsthal, Prince Ernst von col. 122
Heudelet (de Bierre), Étienne, gen. 249, 261
Higgonet, col. 249
Hill, Rowland, gen. 93
Hobe, Leutnant von 189
Hochberg, Wilhem, Baden gen. 153, 154, 155, 279
Hodenberg, Carl von, capt. 119, 120
Hoghton, gen. 78, 85, 86, 87, 89, 267
Hohenlohe, col. von 199, 257
Hohenlohe-Bartenstein, FML Prinz Ludwig von 285

Index of Persons

Hohenlohe-Ingelfingen, Aus. gen. prince 61
Horn, von, col. 290
Houdart-Lamotte, col. 250
la Houssaye, Armand-Lebrun, gen. baron de, 134, 143, 253
Huard, Leonard, gen. 144, 272
Huber, Pierre-François-Antoine, col. 281
Huegel, lt. 119, 120
Hühnerbein, von, gen. 291
Huet, captain 188
Hulin, Pierre-Augustin, gen. 247, 260
Hulot, Étienne, col. 251

Illowaiski v, Cossack, 272
Illowaiski x, Cossack, 296
Illovaiski XII, Cossack gen. 168, 172, 176, 287, 292

Jaquet, col. 272, 288
Jacquinot, Charles-Claude, gen. 231, 289, 298, 299
Jamin, Jean-Baptiste, baron, gen. 288
Jankovich, Rus. gen. 268
Jérôme Bonaparte, King of Westphalia 20, 62, 204, 205
Jett, von, 218, 219, 296
John, Archduke 45
Jomini, Antoine-Henri, gen. 26
Joseph Bonaparte, King of Spain 13, 114
Joubert, Joseph-Antoine-Renée, vicomte de, gen. 288
Junot, Jean-Andoche, gen. 151
Jurschek, Rus. gen. 257

Kachovski, Rus. gen. 71
Kaim, Konrad, Aus. FML 35, 37, 40, 245
Kaminiecki, Polish gen. 167, 283
Kaminski I, Sergei, Rus. gen. 57, 256
Kaminski II, Rus. gen. 258
Kamenskoi, Rus. FM 63, 65, 67, 265
Kamptz, lt. von, 190
Kantakuzen, Rus. gen. 277
Kapsevich, Rus. gen. 134, 143, 276
Karachev, Rus. gen. 296
Katzeler, col. von 186, 291
Kechler, col. von 199, 200
Kellermann, François-Étienne, gen., 24, 31, 33, 37, 38, 41, 42, 43, 44, 97, 167, 227, 233, 244, 249, 254, 299
Kemmis, gen. 77, 90, 267
Kempt, gen. 229
Kienmayer, Michael von, Aus. gen. 53, 59, 100, 255
Kirchner, gen. 163
Kirgener, François-Joseph, col. 252
Kister, Georges, gen. 249, 260
Klebelsberg, FML Graf von 297
Kléber, Jean-Baptiste, gen. 25
Klein, Dominique-Louis-Antoine, gen. 71, 249, 254, 262
Kleist, Friedrich Heinrich Ferdinand Emil, Prus. FM, 168, 177, 214, 223, 285
Klenau, Aus. gen. 168, 170, 173, 177, 285
Klicki, Stanislas, gen. 284
Klöber, von, Prus. officer 181
Klösterlein, gen. von 202
Kluex, von, GM 285
Knesevich, Aus. gen. 245
Knorring, Rus. gen. 73
Koch, von, gen. 296
Kollogribov, Andrei, Rus. gen. 258
Kollowrat, Vincent, Aus. gen. 56, 58, 255, 257
Konopka, Jean, gen. 80, 95
Korf, Rus. artillery officer, 143
Korf, Rus. General-Adjutant 144, 276

Korff, Rus. gen. 75, 264
Korsakov, Rus. lt-gen. 25
Koschin, Rus. gen. 263
Kosen, Rus. artillery col. 144
Krasinski, Polish gen. 283
Kray, Paul von 26, 27
Kretov, Rus. gen. 216, 298
Kreuz, Rus. gen. 276
Kurland, Prinz Biron von 292
Kutaisov, Rus. artillery gen. 134, 277
Kutuzov, Mikhail Larionovich, FM, 50, 51, 53, 122, 123, 124, 125, 129, 146, 150, 156, 197, 255

La Bruyère, André-Adrien-Joseph, gen. 263
Lacour, col. 251
Lacroix, Adjt-Cmdt 253
Lafitte, Michel-Pascal, gen. 283
Lagrange, Joseph, comte, gen. 288, 294
La Houssaye, Armand Lebrun, gen. 128, 275
Lallemand, François-Antoine, gen. 239, 240
La Marseille, Aus. gen. 43
Lambert, Rus. gen. 150
Lamotte, Auguste-Étienne-Marie, gen. 262, 284
Lamsdorf, count 144
Landskoi, Rus. gen. 198
Lange, Gefreiter 190
Langenau, gen. von, 109
Langeron, Louis-Alexandre, gen. 53, 56, 57, 58, 59, 185, 212, 214
Lannes, Jean, marshal, 27, 28, 29, 35, 36, 38, 40, 44, 50, 51, 56, 57, 63, 243, 252
Lapisse, Pierre Bellon, gen. 261
Laplanche, Jean-Baptiste-Antoine, gen. 250, 262
Laplanche-Morthières, Claude-Joseph de, gen. 247
Lapoype, Jean-François Cornu, gen. 32, 33, 244
Lardizabal, Manuel, Spanish gen. 78, 268
Laroche, Baden col. 153, 154, 155
Lasalle, Antoine-Charles-Louis, comte de 67, 70, 250, 263
Lataye, Pierre-François, col. 254
Latham, Matthew, lt. 83, 84
Latour-Maubourg, see Tour-Maubourg
Latrille, Guillaume de, aka Lorencez, col. later comte 251
Lattermann, Christof, Aus. gen. 38, 41, 42, 43
Lauriston, Jacques-Alexandre-Bernard Law, gen. 158, 167, 282
Lavasseur, gen. 251, 261
Laville, Joseph-Alexandre-Felix-Marie, gen. 295
Lavrov, Rus. gen. 277
Lebaron, col. 253
Lecci, Italian gen. 134, 247, 272
Lecourbe, Claude-Jacques, gen. 224
Ledru des Essarts, François-Roche, col. 250, 261
Lefebure, RA officer 88
Lefebvre, col. 250
Lefebvre, François-Joseph, marshal 260
Lefebvre-Desnouëttes, Charles, gen. 203, 232, 239, 299
Lefranc, Jacques, gen. 261
Legende d'Harvesse, François-Marie-Guillaume, col. 252
Legrand, Claude-Juste-Alexandre, gen. 52, 58, 67, 70, 251, 261
Lemonnier-Delafosse, Fr. officer, 115
Leontiev, Rus. gen. 296
Lepel, Westphalian gen. 140, 275

Lepic, Louis, gen. 71
Lessing, von, Saxon gen. 284
Levachov, Rus. col. 143, 287
Leval, Jean-François, gen. 65, 67, 261
Lewis, Rus. gen. 256
Lhuillier de Hoff, François, col. 251
Liallin, Rus. GM, 286
Liechtenstein, Johann, Prinz von, Aus. gen. 56, 57, 255, 257
Liechtenstein, Moritz, Prinz von, Aus. gen. 213, 256
Lieven, Rus. gen. 75, 263
Likhachev, Rus. gen. 134, 141, 276
Lingg, gen. 279
Lippe, von der, lt. 170, 181
Lippert, Hessian lt. 155
Lisanevich, Rus. gen. 168, 287, 297
Lloyd, captain RA, 228
Lobenthal, von, lt-col., 290
Lobau, comte de, see Mouton
Lochet, Pierre-Charles, gen. 74, 249, 260
Löwenstern, Edouard von, ADC 174
Löwenstern, Rus. gen. 277
Loewenwold, Rus. officer, 143
Lorge, Jean-Thomas-Guillaume, gen. 129, 130, 132, 135, 185, 275, 289
Loshakov, Rus. gen. 256
Losthin, von, col. 290
Louis XVIII, king of France 44, 93, 222, 223, 237
Loy, gen. 78, 81, 268
Lubanov, Rus. gen. 259
Lüttichau, Rittmeister von 189
Lützow, Ludwig Adolf Wilhelm von, 195, 199, 200, 201, 205, 206, 292
Lukov, Rus. GM 286
Lumley, Sir William, gen. 78, 81, 88, 267

Macdonald, Jacques-Étienne-Joseph-Alexandre, marshal 157, 165, 183, 212, 213, 214, 215
Mack, Karl, FM, 50
Madden, gen. 78
Mainoni, Joseph-Antoine-Marie-Michel gen. 243
Maison, Nicolas-Joseph, gen. 158, 159, 161, 171, 172, 281, 282
Malachowski, Polish gen. 283
Malher, Jean, gen. 243
Mandeville, Eugène-Charles-Auguste-David, gen. 281, 282
Marchand, Jean-Gabriel, gen. 261
Marcognet, Pierre-Louis Binet de, gen. 229, 230, 232, 298
Marescot, Armand-Samuel, gen. 247
Marie-Louise, Empress 209, 223
Marion, col. 298
Margaron, Pierre, gen. 251, 262
Marizy, Frédéric-Christophe-Henri, gen. 254, 262
Markov, Rus. gen. 64, 67, 174, 264
Marmont, Auguste-Frédéric-Louis Viesse de, marshal 33, 41, 43, 114, 115, 185, 188, 191, 213, 214, 215, 216, 218, 219, 220, 221, 288, 289, 294
Marschalk, von, captain 118
Marwitz, col. von der, 202
Masséna, André, marshal 25, 27, 28, 76
May, John, lt-col. 116
Mazas, col. 250
Mazofskoi, Rus. gen. 263
Mecklenburg, Prinz Karl von 187, 278, 290
Meerheim, Saxon officer 136, 137
Melas, Michael Friedrich Benedikt von, gen. 26, 27, 28, 29, 30, 31, 32, 33, 37, 38, 39, 40, 43, 98, 233, 244
Mellor-Sakomelski II, Rus. gen. 257

Index of Persons

Mensdorff-Poully, col. 203, 292
Mesentziev, Rus. GM 286
Merle, Pierre-Hugues-Victor, gen. 251
Merlin, Antoine-François-Eugène, gen. 281, 289, 295
Metternich, Clemens Wenzel Lothar, Prinz von, Aus. Chancellor 223
Meyers, gen. 81, 88, 89
Michel, Claude-Étienne, count 294
Milet, Jacques-Louis-François, gen. 249, 262
Milhaud, Edouard-Jean-Baptiste, gen. 67, 71, 73, 167, 170, 171, 178, 211, 227, 228, 232, 254, 262, 263, 284, 299
Miliutin, Rus. gen. 259
Miloradovich, Mikhail, Rus. gen. 257, 275
Minkwitz, Rus. gen. 137
Mitzki, Rus. gen. 75
Mohr, Aus. gen. 168, 285
Molard, col. 120
Molitor, Gabriel-Jean-Joseph, gen. 212
Molostov, Rus. adjutant 171
Moltke, col. von 201, 202
Moncey, Bon-Adrien-Jeannot de, marshal 29, 244
Monnier, Jean-Charles, gen. 30, 36, 39, 40, 41, 43, 44, 45, 243
Montbrun, Louis-Pierre, gen. 125, 132, 139, 174, 274
Montelegier, Gabriel-Gaspard-Achille-Adolphe, gen. 284
Montmarie, Aimé-Sulpice-Victor, gen. 283
Moreau, Jean-Claude, gen. 214
Moreau, Jean-Victor, gen. 26, 29, 45, 47
Morand, Charles-Antoine-Louis-Alexis, gen. 134, 138, 250
Morand, Joseph, gen., 196, 260
Morin, Pierre-Nicolas, col. 253
Morio de l'Isle, col. 248
Morland, col., 247
Moroni, col. 272
Mortier, Adolphe-Edouard-Casimir-Joseph, marshal 211, 213, 214, 215, 216, 218, 219, 220, 226, 294
Morzin, Aus. gen. 245
Mourge, sergeant-major 191
Mouton, Georges, gen. comte de Lobau 226
Müller III, Rus. gen. 256
Murat, Joachim, marshal, King of Naples 35, 39, 40, 42, 50, 52, 56, 57, 63, 65, 66, 71, 72, 133, 138, 139, 167, 170, 171, 173, 174, 176, 181, 208, 212, 223, 244, 253, 262, 282
Musnier, Louis, gen. 243
Mutel, porte-aigle, 191
Mutius, Prus. col. 158, 159, 161, 172, 281, 286
Myer, gen. 88, 267

Nansouty, Étienne-Marie-Antoine, gen. 126, 174, 253, 262
Napoleon Bonaparte, Emperor 12, 13, 16, 18, 19, 22, 24, 25, 26, 27, 28, 29, 30, 32, 33, 35, 36, 37, 39, 40, 41, 43, 44, 46, 47, 48, 49, 50, 51, 52, 56, 58, 59, 60, 61, 62, 63, 64, 65, 66, 67, 70, 71, 72, 73, 74, 75, 87, 93, 95, 100, 101, 114, 122, 123, 124, 125, 136, 137, 139, 140, 141, 145, 146, 150, 151, 152, 155, 156, 157, 158, 162, 164, 165, 166, 171, 173, 183, 185, 194, 195, 196, 197, 207, 208, 209, 210, 211, 212, 213, 214, 215, 219, 220, 221, 222, 223, 224, 225, 226, 227, 232, 233, 243, 247, 260
Narboni, col. 272
Nelson, Horatio, vice-admiral 24
Nerin, col. 252
Netting, Rus. gen. 263

Neverovski, gen. 278
Ney, Michel, marshal 63, 64, 65, 67, 73, 152, 159, 162, 214, 221, 225, 227, 232, 237, 261
Nicolas, Jean, col. 252
Nikitin, Rus. gen. 134, 173
Nimbsch, Aus. gen. 245
Nobili, Aus. gen. 38, 245
Noirot, Jean-Baptiste, col. 254
Normann, gen. Graf von 186, 199, 200, 201, 202, 289
Nostitz, Aus. gen. 216, 218, 264, 297

Ochs, gen. von 199
Odeleben, col. Freiherr 162
O'Donell, Graf, Aus. gen. 19
Örtzen, Victor von 190
Olsuviev, Rus. gen. 212
Orange, Willem, Prince of 225
Ordener, Michel, gen. 247
O'Reilly, Andreas, Graf von, Aus. gen. 30, 32, 33, 34, 37, 38, 39, 42
Orlov-Denisov, count, col. 292, 296
Ornano, Philippe-Antoine, gen. 126, 128, 273
Osten-Sacken, (often referred to as Sacken) Fabian Gottlieb, gen. Graf von der 67, 185, 211, 212, 214, 263
Ostermann-Tolstoi, Rus. gen. 63, 67, 126, 141, 263, 275
Ott, Karl von, Aus. FML 30, 33, 35, 36, 38, 39, 43, 246
Otway, gen. 78, 268
Oudinot, Nicolas-Charles, marshal 151, 152, 185, 213
Outechine, Rus. corporal 71

Paar, Aus. grenadier commander 42
Pack, Sir Dennis, gen. 229
Pacthod, Michel-Marie, gen. 219, 220, 248, 295
Pages, Joseph, col. 251
Pahlen II, Pavel Petrovich, Rus. lt-gen. 168, 174, 175, 176, 177, 287
Pahlen III, Piotr Petrovich, Rus. lt-gen. 297
Pahlen, Ivan, count 176, 265
Pahlen, Magnus, gen. 203, 293
Pahlen, Piotr Piotrovich, Rus. gen. count 150, 156, 216, 218, 219, 263, 276
Pajol, Claude-Pierre, gen. 167, 168, 176, 284
Palota, Johann Frimont von, col. 244
Panchulidsev, Rus. gen. 144
Partouneaux, Louis, gen. 152, 279
Paskevich, Rus. gen. 277
Paskevich, Ivan Federovich, Rus. FM 134
Pastol, gen. 273
Paul I, Czar of Russia 16
Paumgarten, Aus. gen. 172, 285
Pelleport, Pierre, gen. 191, 288
Penitzki, Rus. gen. 258
Penne Villemur, Spanish gen. 78, 269
Peraldi, col. 272
Pescheux, col. 248
Petit, author 37
Philippe, captain, 117
Philippon, Armand, col. 248
Phillips, captain, 85
Picard, Joseph-Denis, gen. 254, 262
Picquet, Cyrille-Simon, gen. 283, 294
Picton, Sir Thomas, gen. 229
Pilatti, Giovanni, Aus. gen. 37, 42
Pinot, Italian gen. 273
Pirch I, von, Prus. col. 159, 281, 285
Piré, Hippolyte-Marie-Guillaume de Rosnyvinen, gen. 240, 241, 283
Piston, Joseph, gen. 253

Pitt, William 46
Planzeaux, col. 281
Platov, Matvei Ivanovich, Ataman, 102, 122, 123, 125, 126, 128, 129, 145, 203, 272, 293, 296
Podvoronti, Rus. dragoon 66
Poniatowski, Prince Josef Anton, marshal, 167, 172, 283
Ponsonby, Frederick, lt-col. 232
Ponsonby, Sir William, gen. 227, 229, 232, 233, 299
Poser, von, Prus. officer 180
Post, trooper, 117
Pouget, François-René, col. 251
Pouzet, col. 250
Preussen, August Prinz von, GM 286
Preval, Claude-Antoine-Hippolyte col. 253
Preysing, von, gen. 273
Przbychevski, Ishami, Rus. gen. 256
Puckler, count 292
Puschnitzki, Rus. GM 287

Quenuot, Mathieu, gen. 284

Radionov, Rus. Cossack, 287
Radziwill, Michel, Polish gen. 279
Ravier, Jean-Baptiste-Ambroise, col. 251
Raevski, Rus. gen. 123, 124, 126, 128, 129, 132, 134, 136, 137, 138, 139, 141, 144, 277
Rapp, Jean, gen. 100, 224
Razout, Jean-Nicolas, col. 248
Rembow, von, gen. 265
Rechberg, gen. count von 198
Reille, Honoré-Charles-Michel-Joseph, gen. 226
Reinhard, lt-col. von, 219
Reitzenstein, capt. von, 117, 119, 120
Repnin, Rus. gen. prince 59
Repninski II, Sergei, Rus. gen. 257
Rettberg, capt. 229
Retz, Aus. gen. 246
Rey, Jean-Pierre-Antoine, col. 251
Rheinhold, Gefreiter 189
Ricard, Étiennee-Pierre-Sylvestre, gen. 294
Richemont, Christophee-François, gen. 288
Richter, Jean-Louis, gen. 274
Rieu, Jean-Louis, captain 190
Rigeaud, col. 250, 295
Rivaud, Jean, gen. 36, 38, 43, 244
Rivaud de la Raffinière, Olivier Macoux, gen. 243, 248
Rochambeau, Donatien-Marie-Joseph de Vimeur, gen. 283
Röder, Prus. gen. 168, 170, 177, 286
Roget de Belloguet, Mansuy-Dominique, gen. 253, 262
Rogoyski, Lieutenant 80, 95
Roguet, François, gen. 129, 261
Romaña, La, Spanish gen. 91
Rosen, Rus. gen. 296
Ross, captain RA 228
Roth, Rus. GM 286
Roth von Schreckenstein, gen. Freiherr 129, 133, 137, 140, 143
Rottermund, Rus. gen. 257
Roussel, Jean-Claude, gen. 272
Roussel, François-Xavier, gen. 247
Roussel d'Hurbal, Nicolas-François, gen. 295, 299
Rousseau, Aus. gen. 244
Rouvroy, col. 288
Rouyer, Jean-Victor, gen. 64
Rozniecki, Polish gen. 141, 275
Rüdiger, Rus. gen. 168, 172, 287, 297
Ruffin, François-Amable, gen. 248

Sacken, Rus. gen. 264

Index of Persons

Sahuc, Louis-Michel-Antoine, gen. 250, 262
St. Dizier, col. 250
St. Genies, Jean-Marie-Noel Delisle de Falcon, gen. 274
St. Germain, Antoine-Louis Decrest, gen. de 253
St. Hilaire, Louis-Vincent-Joseph le Blonde, gen. 67, 70, 71, 250, 260
St. Julien, Franz, Aus. gen. 41, 42, 245
St. Priest, Rus. gen. 214
St. Remond, col. 249
St. Sulpice, Raymond-Gaspard de Bonardi, gen. 254, 262
Saligny, Charles, gen. 250
Salis, von, Aus. col. 171
Sass, Rus. dragoon col. 144
Savary, Anne-Jean, 40, 247
Saxe-Weimar, gen., Prinz von 226
Scalfort, Nicolas-Joseph, gen. 251
Schachafskoi, Rus. GM 287
Schack, major von 188
Schaeffer, von, Aus. GM 285
Schalnitzki, Rus. col. 287
Scharnhorst, Gerhard Johann David von, gen. 74
Schellenberg, Aus. FML. 38, 39, 40, 246
Schilt, Jean, gen. 243
Schiner, Joseph-François-Ignace-Maximilien, gen. 250, 261
Schischkin, Rus. major 176
Schmidt, Aus. gen. 52
Schmidt, von, lt-col. 291
Schobert, Laurent, col. 251
Schüssler, lt. 189
Schütz, Baden Feldjäger 153
Schufanov, Rus. col. 168, 287
Schustekh, Aus. col. 38, 39
Schwartz, François-Xavier de, col. 254
Schwarzenberg, Karl Philipp Fürst zu, FM 109, 165, 172, 183, 211, 212, 213, 215, 224, 285
Sebastiani, Adjt-Cmdt 253
Sebastiani de la Porta, Horace-François-Bastian, gen. 95, 135, 253, 274
Senerols, gen. 249
Senft von Pilsach, Prus. officer 179
Seron, gen. 275
Seslavin, Rus. hero 175, 212, 293, 296
Seyboldsdorff, col. von 197
Seydewitz, gen. von, 273
Seydlitz, von, Prus. gen. 173
Shaw, John, corporal 28
Sherer, Moyle 86
Sierawski, Polish gen. 283
Sievers I, gen. 278
Simmer, gen. 282
Sivray, de, gen. 273
Smith, Captain Sir William Sidney 25
Sohr, major von 186
Sokolniki, Polish gen. 167
Somerset, Lord Edward, gen. 227, 228, 229, 231, 299
Somov, Rus. gen. 67, 70, 264
Songis des Corbins, Nicolas-Marie, gen. 249
Sopransi, Louis-Charles-Barthelemy, gen. 284
Sorbier, Jean-Barthelemy, gen. 249
Soules, Jérôme, gen. 247, 260
Soult, Nicolas Jean-de-Dieu, marshal 51, 52, 53, 56, 57, 59, 65, 66, 70, 77, 78, 80, 81, 87, 88, 89, 90, 91, 92, 93, 102, 208, 250, 260, 269
Sparre, Louis-Ernest-Joseph, gen. 295
Spedding, captain, 85
Splenyi de Milhaldy, Aus. gen. 172, 285
Springer, Martin, sgt-major 154, 155

Steinmetz, von, col. 290
Stevens, William, captain 83
Stewart, Sir Charles, gen. 187
Stewart, Sir William, gen. 78, 81, 82, 85, 87, 90, 92, 267
Sticker, Aus. gen. 40, 246
Stiles, corporal 230
Strik, Rus. gen. 256
Stutterheim, Aus. gen. 264
Subervie, Jacques-Gervais, gen. 132, 168, 170, 171, 274, 284
Suchet, Louis-Gabriel, marshal 30, 56, 57, 91, 208, 224, 252
Sukin II, Rus. gen. 75, 263
Suvorov, Alexei Vasilievich, Rus. FM, 25

Talleyrand, Charles-Maurice 220
Tappe, cornet 119, 120
Tauentzien, Bolesas Friedrich Emanuel, gen. 183
Tettenborn, Rus. col. 20, 197, 206, 293
Thiebault, Paul-Charles-François, gen. 57, 250
Thielemann (or Thielmann), von, gen. 130, 135, 137, 139, 141, 143, 203, 226, 275, 292
Thiry, Nicolas-Marin, gen. 275, 295
Thomas, Edward Price, ensign 83
Thomières, Jean-Guillaume-Barthelemy, gen. 114
Thuillier, col. 205, 206
Timm, Mecklenburg officer 187, 190
Titov II, Rus. gen. 75, 263
Toli, François, double agent 29
Toll, col. (later gen.) von 122, 123, 125, 173
Tour-Maubourg, Marie-Victor-Nicolas de Fay, marquis de la, gen. 80, 81, 82, 83, 87, 88, 90, 126, 255, 129, 130, 133, 135, 137, 138, 139, 140, 141, 174, 262, 270, 275, 283
Travers, Étienne-Jacques, gen. 228
Trelliard, Anne-François-Charles, gen. 254
Trenck, von, GM 297
Triaire, gen. 272
Turno, Polish gen. 275
Turreau, Louis-Marie, gen. 27
Tutchkov I, Rus. gen. 67, 73, 264

Ulm, Aus. gen. 39
Uminski, Polish cavalry officer, 283
Urusov, Alexander, Rus. gen. 256
Uschakov, Rus. gen. 263
Uslar, Friedrich von, 118, 120
Uvarov, Fedor, Rus. gen. 103, 123, 124, 125, 126, 128, 129, 145, 272
Uxbridge, Lord, Henry, gen. 231, 233

Valentini, lt-col., Freiherr von 289
Valhubert, Jean-Marie-Mellon-Roger, gen. 252
Vallin, Louis, gen. 281, 284
Valory, Guy-Louis-Henri, gen. 282
Vandamme, Dominique-Joseph-Renée, gen. 58, 59, 250
Vandeleur, Sir John, gen. 232
Varé, Louis-Prix, gen. 74, 251, 260
Vasilchikov, Rus. gen. 277
Vedel, Dominique-Honoré-Antoine-Marie, col. comte de 252
Verdier, Jean-Christophe-Collin, gen. 250, 262
Vial, Honoré, gen. 282
Vial, Jacques-Laurent-Louis-Augustin, gen. 284
Viallanes, Jean-Baptiste-Theodore, gen. 249, 260
Victor (Perrin), Claude, marshal 32, 35, 38, 39, 146, 147, 152, 153, 154, 155, 167, 243, 279, 282

Villata, Italian gen. 273
Villatte, Eugene-Casimir, gen. 261
Villeneuve, Pierre-Charles -Jean-Baptiste-Sylvestre, vice-admiral 49
Vincent, Henri-Catherine-Baltazard, col. 281
Vivies, Guillaume-Raymond-Amant, gen. 250, 261
Vlasov III, Cossack commander, 272
Vlastov, Rus. gen. 153, 281, 286
Vodnianski, Rus. gen. 257
Vogelsang, Ludwig von, Aus. gen. 43, 246
Volkonski, Gen Adjt, prince 255
Voronzov, Rus. gen. 214, 278
Voropaitski, Rus. gen. 258
Voss, lt. 119, 120
Vukassovich, Aus. FML 29
Vustov, Rus. GM 287

Wahlen-Jürgass, Freiherr von, col. 192, 281, 291
Walsh, Charles, ensign 83, 84
Walther, col. 252
Walther, Frédéric-Henri, gen. 253, 254, 260
Warburg, lt-col. 189, 191
Walsleben, von, gen. 296
Wathiez, François-Isidore, gen. 295
Watier, gen. 133, 135, 138, 141, 144, 263, 274, 299
Watrin, François, gen. 35, 38, 243
Weber, Aus. gen. 257
Wedel, Erhard-Gustave, gen. 63
Weidenfeld, Aus. gen. 37, 43, 245
Wellington, Arthur Wellesley, FM Duke of 76, 77, 92, 93, 114, 115, 116, 120, 208, 209, 223, 224, 225, 226, 227, 231, 232, 237
Welzien, von, col. 290
Werder, von, major 132
Werle, François-Jean, gen. 80, 81, 87, 88, 89, 248, 269
Weyrother, Aus. gen. 52, 255
Wilhelm, Elector of Hesse-Kassel 205
Winzingerode, Gen Adjt 255
Winzingerode, Ferdinand, gen. 50, 197, 203, 213, 214, 215
Wittgenstein, Ludwig Adolf Peter, FM count 147, 150, 152, 153, 158, 167, 168, 170, 177, 212, 258, 280, 286
Wojciechowski, lt. 80, 95
Woltersdorf, Gefreiter 189
Wrede, von, Bavarian gen. 211, 249
Württemberg, Adam, prince, gen. 216, 219, 296
Württemberg, Eugen von, gen. duke, 126, 168, 287
Württemberg, Wilhelm, Crown Prince, gen. 216, 218, 296

Yorck, David Ludwig von, FM 157, 166, 183, 185, 186, 187, 188, 192, 211, 212, 213, 214, 289
Yvendorf, Jean-Frédéric, col. 253

Zach, Anton von, Aus. gen. 31, 40, 41, 42, 98, 244
Zandt, von, gen. 204
Zapolski, Rus. gen. 265
Zayas, José, Spanish gen. 78, 81, 82, 84, 85, 90, 268
Zeppelin, von, col. 290
Zhirov, Cossack, 272
Zielinski, col. 289
Zieten, Hans Ernst Karl von, Prus. FM 159, 225, 281, 286